# Contents

# Music and Dementia

# Music and Dementia

## From Cognition to Therapy

*Edited by*

AMEE BAIRD

SANDRA GARRIDO

JEANETTE TAMPLIN

OXFORD

UNIVERSITY PRESS

# OXFORD
UNIVERSITY PRESS

Oxford University Press is a department of the University of Oxford. It furthers the University's objective of excellence in research, scholarship, and education by publishing worldwide. Oxford is a registered trade mark of Oxford University Press in the UK and certain other countries.

Published in the United States of America by Oxford University Press
198 Madison Avenue, New York, NY 10016, United States of America.

© Oxford University Press 2020

All rights reserved. No part of this publication may be reproduced, stored in a retrieval system, or transmitted, in any form or by any means, without the prior permission in writing of Oxford University Press, or as expressly permitted by law, by license, or under terms agreed with the appropriate reproduction rights organization. Inquiries concerning reproduction outside the scope of the above should be sent to the Rights Department, Oxford University Press, at the address above.

You must not circulate this work in any other form
and you must impose this same condition on any acquirer.

CIP data is on file at the Library of Congress
ISBN 978–0–19–007593–4

1 3 5 7 9 8 6 4 2

Printed by Integrated Books International, United States of America

# Preface

The global statistics of increasing rates of dementia in our aging population are well known and alarming. Dementia is an umbrella term for a group of neurological conditions that cause a gradual death of brain cells and associated decline in cognitive functions that lead to significant impairment in everyday functioning (American Psychiatric Association, 2013; Camicioli, 2014). There are various types and causes of dementia, the most common being Alzheimer's Dementia (AD) caused by Alzheimer's disease (Alzheimer's Association, 2018). It is clear that dementia will be a leading cause of disability in the near future, with associated significant financial and societal burdens. With no cure in sight, there is an urgent need for the development of treatments to alleviate symptoms of dementia and to improve quality of life for people living with a diagnosis. Numerous pharmacological treatments exist, but their limited efficacy has prompted the growing need for the development of nonpharmacological treatment options (Galik, 2016; National Institute for Health and Care Excellence, 2018). There is accumulating evidence that music is one of the most effective nonpharmacological interventions (Chang et al., 2015), and it is for this reason that the current volume—which collates the available evidence on the topic music and dementia—is timely.

The vast majority of research on music and dementia has focused on people with AD, which is understandable given that it is the most common type of dementia. The hallmark symptom of AD is impaired memory function, and the formal diagnostic criteria require that there is impairment in at least one other cognitive domain (such as language skills) in addition to memory (McKhann et al., 2011). In the early stage, the neuropathological effects of AD are focal and typically arise in the temporal lobes, with more widespread effects across other brain regions as the disease progresses. Currently there is relatively limited research on the effects of music on people with non-AD types of dementia, and the work that has been done is primarily focused on frontotemporal dementia (FTD). Currently, FTD is the second most common type of dementia and usually has an earlier age of onset compared with AD (45–60 years compared with over 65 years in

AD). There are two main variants of FTD, each with their own diagnostic criteria: (a) Behavioral variant FTD (Bv-FTD) causes prominent changes in social and emotional functions, such as disinhibition (Rascovsky et al., 2011) and (b) Primary Progressive Aphasia (PPA), which has three subtypes, namely Semantic Dementia (SD), characterized by impaired naming and comprehension abilities (Landin-Romero, Tan, Hodges, & Kumfor, 2016), Nonfluent/Agrammatic PPA, with the main feature of effortful/halting speech with agrammatism of language production, and Lopogenic PPA, which causes impaired single-word retrieval in spontaneous speech and im-paired repetition of sentences/phrases (Gorno-Tempini et al., 2011). The neuropathology of FTD primarily affects frontal and anterior temporal brain regions, with each variant having distinct neuropathological and genetic profiles.

As editors we come from different disciplines and currently hold Dementia Research Development Fellowships co-funded by the Australian National Health and Medical Research Council and the Australian Research Council. It was our common interest in music and dementia along with our diverse disciplinary perspectives, namely clinical neuropsychology (Amee Baird), music psychology (Sandra Garrido), and music therapy (Jeanette Tamplin), that inspired us to work together on this interdisciplinary volume. The authors of each chapter are international experts from our three disciplines working in the field of music and dementia. Each of us are musicians and have had a long-standing interest in music and well-being. Amee Baird has worked as a clinical neuropsychologist for 15 years. Her interest in the connection between music and dementia was inspired during her postdoc-toral research position focusing on the neuropsychology of music memory, during which she read literature about preserved music memory and skills in people with severe AD. She has spent over a decade undertaking research in this field and has a particular interest in music-evoked autobiographical memories and preserved music skills in the face of cognitive decline, and how this phenomenon can enhance the sense of self in people with dementia. Sandra Garrido has been researching in the field of applied music psychology for more than 10 years. Music psychology is concerned with the cognitive, behavioral, and emotional processes involved in responding to and making music. Her interest in this field comes from her experience in working as a music educator when she became interested in the effects of engaging with music on the mental health of her students. Her research since then has fo-cused on the relationship between depression and music in both adolescents and older people with dementia. Jeanette Tamplin has been working as a registered music therapist in neurorehabilitation for 20 years. In her clinical work with people with dementia she has seen how music can be used to level the playing field between people with a diagnosis and their healthy peers and loved ones. This is based on her experience of music as a strength and resource for people living with dementia. She is also a strong advocate for the benefits of music therapy as an effective nonpharmacological intervention that can ameliorate psychological and behavioral challenges often experienced by people with dementia (sometimes described as 'behavioural and psychological symptoms of dementia' or BPSD) through enhancing personhood and providing engaging and stimulating experiences.

These differing disciplinary approaches consider the subject of music and dementia at numerous levels—the biological, the behavioral and cognitive, and the phenomenological. Thus, the current volume attempts to draw together these three perspectives in the material covered herein, presenting a comprehensive and multidisciplinary review of cutting-edge research in this area. This book is organized into three broad parts, each incorporating our three disciplines.

Part I focuses on the question, "Why Music for People With Dementia?" and contains an introductory chapter from authors in each of the three disciplines. In Chapter 1, Clements-Cortés addresses this question from a music therapy perspective and explores different music engagement opportunities for people with dementia. The continuum of musical experiences outlined ranges from informal music listening experiences, to community music participation opportunities, to more supportive structures offered in formal music therapy. In Chapter 2, Ghilain, Schiaratura, Singh, Lesaffre, and Samson explore the question from a neuropsychological perspective, exploring the unique effect of music in dementia with a focus on the topic of entrainment. In Chapter 3, Brancatisano and Thompson consider the issue from a music psychology point of view and identify seven capacities of music that make it an ideal treatment tool for people with dementia.

Part II explores the "Impact of Music on Cognition and Emotion in People With Dementia" from the perspectives of neuropsychology and music psychology. In Chapter 4, Vanstone and Cuddy focus on music engagement from the basis of preserved musical memory in people with AD. Chapter 5 provides a review by Omar of music cognition in people with non-AD dementias, specifically variants of FTD. In Chapter 6, Särkämö reviews how various music activities can facilitate cognitive (including memory, language, attention, and executive functions) and emotional functions (in particular reducing depression symptoms), and enhance well-being in people with dementia. Garrido explores the increasingly popular use of personalized musical playlists for people living with dementia and its impact on depression in Chapter 7. In Chapter 8, Baird and Thompson review musical instrument playing in people with dementia, and how this can access preserved forms of memory and the sense of self.

Part III contains five chapters that focus on "Music Therapy in Dementia Care." This section covers music therapy research and practice across the spectrum of dementia severity from community-based therapeutic programs to the advanced stages of the disease. In Chapter 9, Dowson and McDermott describe approaches to measuring the impact of music therapy and music activities in dementia care. Both quantitative and qualitative approaches to measurement are presented with a discussion of the benefits and limitations of each. Chapter 10 presents an umbrella review and recommendations by Lipe and Edmonston for music and music therapy interventions targeting the reduction of BPSD. In Chapter 11, Ridder and Ørnholt Bøtker describe music therapy approaches in advanced dementia care and discuss options for skill sharing between music therapists and care staff to enhance well-being for both caregivers and care recipients. Mercadel-Brotons presents a narrative synthesis of music therapy interventions for advanced dementia in Chapter 12. In Chapter 13, Tamplin and Clark explore therapeutic music interventions to support people with dementia living at home with their family caregivers, with a specific focus on communal singing and therapeutic songwriting interventions. This chapter also outlines the differences in services provided by music therapists and community musicians, and highlights opportunities for caregivers to harness the therapeutic music affordances of music with their loved one with dementia in the home environment. In the concluding chapter (Chapter 14), the editors suggest future directions for researchers and practitioners working with music in dementia care.

At the time of publication, funding of over two million euros from the Joint Programme for Neurodegenerative Diseases was announced for a new large international randomized controlled trial of a home-based family caregiver-delivered music intervention for people living with dementia (HOMESIDE). The study will evaluate the effect of training 500 caregivers (from five countries—Australia, UK, Germany, Norway, and Poland) in how to best use musical interventions with their family member who has dementia. This substantial commitment of global funding to investigate the therapeutic benefits of music in dementia care further ensures the sustainability and accessibility of music for people with dementia. It demonstrates that the strong impact of music in dementia is already recognized, as this seminal book testifies.

The research that has been compiled in this volume demonstrates the vast potential that music has for improving the quality of life of people with dementia. It is our hope that this compilation of current research on music and dementia will not only provide a useful resource for healthcare professionals, students, carers, and anyone with an interest in this field, but also inspiration for further interdisciplinary research in this area.

## References

Alzheimer's Association. (2018). 2018 Alzheimer's disease facts and figures. *Alzheimer's Dementia, 14*(3), 367–429.

American Psychiatric Association. (2013). *Diagnostic and statistical manual of mental disorders* (5th ed.). Arlington, VA: American Psychiatric Publishing.

Camicioli, R. (2014). Diagnosis and differential diagnosis of dementia. In J F. Quinn (Ed.), *Dementia* (pp. 1–14). Chichester, England: Wiley.

Chang, Y.-S., Chu, H., Yang, C.-Y., Tsai, J.-C., Chung, M.-H., Liao, Y.-M., . . . Chou, K.-R. (2015). The efficacy of music therapy for people with dementia: A meta-analysis of randomised controlled trials. *Journal of Clinical Nursing, 24*(23–24), 3425–3440. doi:10.1111/jocn.12976

Galik, E. (2016). Treatment of dementia: Non-pharmacological approaches. In M. Boltz & J. E. Galvin (Eds.), *Dementia care: An evidence-based approach* (pp 97–112). Switzerland: Springer.

National Institute for Health and Care Excellence. (2018). *Dementia: Assessment, management and support for people living with dementia and their carers.* Retrieved from https://www.nice.org.uk/guidance/ng97

Rascovsky, K., Hodges, J. R., Knopman, D., Mendez, M. F., Kramer, J. H., Neuhaus, J., . . . Hillis, A. E. (2011). Sensitivity of revised diagnostic criteria for the behavioural variant of frontotemporal dementia. *Brain, 134*(9), 2456–2477.

# Contributors

**Amee Baird, PhD, MPsych** (Clinical Neuropsychology)
Macquarie University & Newcastle Neuropsychology, Australia

**Julie Ørnholt Bøtker, MA, DMTF**
Farsøhthus Plejecenter, Nordjurs, Denmark

**Olivia Brancatisano**
Macquarie University, Australia

**Imogen N. Clark, PhD, RMT**
The University of Melbourne, Australia

**Amy Clements-Cortés, PhD, RP, MT-BC, MTA**
University of Toronto, Canada

**Lola L. Cuddy, PhD**
Queen's University, Canada

**Becky Dowson, MA**
University of Nottingham, UK

**Molly Edmonston, MMT, CDP**
The Village at Orchard Ridge, Winchester, USA

**Sandra Garrido, PhD**
Western Sydney University, Australia

**Matthieu Ghilain, MA**
Université de Lille, Villeneuve d'Ascq, France

**Micheline Lesaffre, PhD**
Ghent University, Belgium

**Anne W. Lipe, PhD, MT-BC**
Shenandoah University, USA

**Orii McDermott, PhD**
University of Nottingham, UK

**Melissa Mercadal-Brotons, PhD, MT-BC, SMTAE**
Escola Superior de Música de Catalunya, Spain

**Rohani Omar, MD(Res)(UCL), MA(Cantab), MRCP(UK), DipABRSM**
University College London, UK

**Hanne Mette Ridder, PhD, DMTF**
Aalborg University, Denmark

**Séverine Samson, PhD**
University of Lille, Villeneuve d'Ascq, France & Pitié-Salpêtrière- Charles Foix, Paris, France

**Teppo Särkämö, MA, PhD**
University of Helsinki, Finland

**Loris Schiaratura, PhD**
University of Lille, Villeneuve d'Ascq, France

**Ashmita Singh, BMSc**
Western University, London, Ontario, Canada

**Jeanette Tamplin, PhD, RMT**
The University of Melbourne, Australia

**William Forde Thompson, PhD**
Macquarie University, Australia

**Ashley D. Vanstone, PhD**
Avon and Wiltshire Mental Health Partnership NHS Trust, UK

# PART I
# WHY MUSIC FOR PEOPLE WITH DEMENTIA?

# 1

# Understanding the Continuum of Musical Experiences for People With Dementia

*Amy Clements-Cortés*

## Introduction

At the present time with technologic advancements, music is more readily available than in any other period in history. Our society has also experienced a shift in the services provided by healthcare professionals in the past 30 years, moving away from the medical model of health and illness focusing on a physical and curative level, to focusing on more holistic, preventative healthcare and health promotion. Increasingly, more people are interested in exploring alternative and complementary methods of addressing health concerns. These holistic models suggest a more comprehensive evaluation of the vulnerable individual in medical care, which involves not only the physical aspect of well-being, but also the social, emotional, and spiritual domains. Included in these alternative or complementary methods are a variety of musical experiences which will be explored in this chapter, with a focus on their benefit for people with dementia.

Musical experiences exist on a continuum of recreational to therapeutic opportunities, allowing for both instances (such as recreational music making) where both therapeutic and recreational aims are targeted simultaneously (see Figure 1.1). Further, musical experiences can be receptive or active in nature. Receptive experiences include music listening and music for relaxation, while active music making comprises experiences such as playing instruments and moving to music. Although there is immense value in all musical experiences, it is important to understand the differences between them in order to accommodate the unique needs of the vulnerable individual. Specifically, when a person with dementia is unable to make decisions for him/herself, understanding the difference between the range of musical

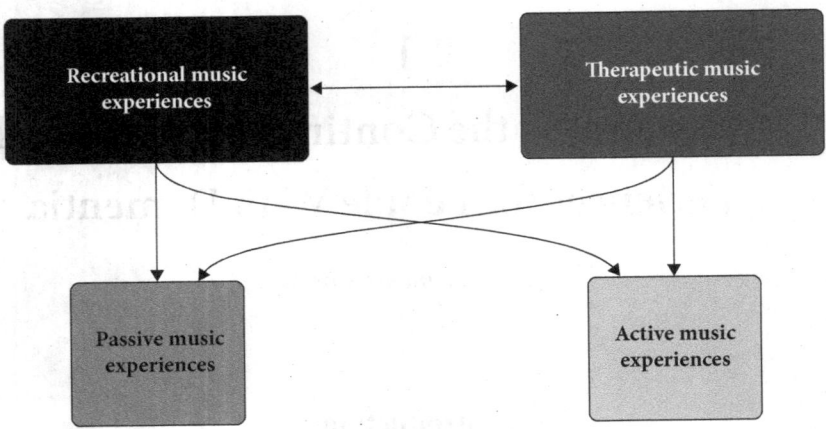

**Figure 1.1.** Continuum of music experiences.

experiences available may guide caregivers in selecting the most appropriate, safe, and beneficial interventions for the individual.

## A Starting Point: Using Accurate Terminology

There is considerable confusion regarding the identification and terminology of musical experiences in healthcare settings. In this chapter, these musical experiences will be clearly defined, providing practical information to enable selection of the most appropriate experiences when working with people with dementia. Frequently, many music experiences in medical settings are categorized as "music therapy" (Garrido et al., 2017). Unfortunately, this description is not always accurate. Defining all music experiences as music therapy can convey incorrect information regarding the clinical practice of music therapy, and particularly the importance of clinical training and experience provided by a qualified music therapist. Yinger and Gooding (2015) acknowledge that there is a need for clearer distinction between music medicine and music-therapy interventions. In 2014, the American Music Therapy Association (AMTA) highlighted some examples of experiences that are not music therapy. These illustrations provide helpful guidance in understanding the types of musical experiences that are often mislabeled as "music therapy."

- A person with Alzheimer's listening to an iPod with headphones of his/ her favorite songs
- Groups such as Bedside Musicians, Musicians on Call, Music Practitioners, Sound Healers, and Music Thanatologists
- Celebrities performing at hospitals and/or schools
- A piano player in the lobby of a hospital
- Nurses playing background music for patients
- Artists in residence
- Arts educators
- A high school student playing guitar in a nursing home
- A choir singing on the pediatric floor of a hospital (Cisin, 2014)

This chapter will provide an overview of the benefits of many types of music opportunities for people with dementia and a discussion of possible considerations for using music with this population. Basic definitions of the various types of music experiences along a continuum will be presented, alongside a list of implications for care providers.

## Overview of Music Benefits for People With Dementia

Research has shown numerous benefits of engagement with various musical experiences for people across the lifespan, including individuals diagnosed with a cognitive impairment or dementia. These benefits may offer outcomes displayed in a number of domains including physical, psychological, cognitive, communicative, emotional, social, and spiritual. For example, a number of researchers have found that music had a positive impact on behaviors that were commonly exhibited by people with dementia including irritability, aggression, wandering, repetitive behaviors, and poor attention spans (Ledger & Baker, 2007; Groene, 1993; Jennings & Vance, 2002; Koger & Brotons, 2000; Raglio et al., 2008). With respect to music therapy, several authors have noted that utilizing music from the young adult years of a person with dementia were most significant in improving cognitive abilities of this population (Bruer, Spitznagel, & Cloninger, 2007; Van de Winckel, Feys, De Weerdt, & Dom, 2004). Raglio (2010) reported the potential for music therapy to activate an innovative dialogue with a person with dementia, particularly creating a relational-emotional dimension, not achieved in other modalities. Similarly, Koger and Brotons (2000) attributed music therapy to the improvement of speech fluency and content. Sixsmith and Gibson (2007) noted that even though cognitive deterioration is a primary symptom of dementia, the capacity to engage with music often remains. Within each of the musical experiences outlined in this chapter you will find examples of these types of benefits. It is important to note there are times when music therapy may provide the most optimal benefits and is indicated above the inclusion of all other musical experiences (Garrido et al., 2017) or in collaborative treatment including other musical experiences. There are other scenarios where the expertise of a music therapist may not be required to facilitate musical experiences for a person with dementia. For example, research in mental health suggests that there is not a universal benefit of music on mood. For this reason, when considering music-based experiences for vulnerable people with dementia, it is imperative to consider when a music therapist may be required to process emotions stimulated by music in a supportive and safe environment (Bunt & Hoskyns, 2013).

## Music Considerations

An important consideration alongside the various forms of musical experiences is the type of music that will be used for the intervention. Both new and familiar music have a place in dementia care depending on the needs of the individual. For people with dementia, preferred music is often a valuable tool in assisting a person to connect to their identity and history (McDermott, Orrell, & Mette Ridder, 2014), as it provides a springboard for reminiscence. When a provider is responsible for selecting music for a person with dementia, it is important to consider possible past associations to the music selection. For example, if a song is linked to a negative memory of war, music of this nature should be implemented by a music therapist, in order to help with the therapeutic processing of emotions that may arise through the experience. Furthermore, there is the possibility that a previous positive association with a song from the past may no longer retain that same sentiment in the context of a different auditory experience. Hogan (1998) states that preferred music can foster creative participation, enhance a person's self-worth and sense of accomplishment, stimulate their mind and body, and connect to their emotional needs. Determining music preferences for a person with dementia can be challenging, however. While a good starting point is to ask the person and their family members what music the person has enjoyed, it is not always possible to ascertain this information. It may help to look to popular music and artists from the time when a person with dementia was in their young adult years, or to cultural music or music that was part of their upbringing. However, although likely to be familiar, it cannot be assumed that this is the person's preferred music.

Listening to preferred music also plays a role in pain perception, and while not specifically related to dementia, pain is a common symptom in older adults. Mitchell and Macdonald (2006) refer to preferred music as a medium for modulating pain tolerance and control; it also serves as a distraction and has a positive affect impact. It is important to note that Garrido et al. (2017) state that "there has been relatively little empirical study of the effectiveness of nontherapist led playlist interventions in comparison to music therapy" (p. 1130), and Aldridge (1994) acknowledged the evidence that music therapy plays a significant role in improving quality of life of people living with Alzheimer's disease.

Depending on the severity and complexity of dementia, simpler music that includes fewer instruments or simpler musical elements may be beneficial in order to reduce the potential for sensory overload. The piano can be an ideal music instrument for accompanying and supporting musical experiences for people with dementia, due to its familiarity and preference for many older adults (Clements-Cortés, 2014a). The guitar is also a practical and advantageous accompanying instrument, as it allows the facilitator to be in close proximity to the persons with dementia, is portable, and works in smaller spaces and in individual sessions held in a person's room. The ukulele might also be used to accompany in small groups or individual sessions. Small percussion instruments are also useful to engage persons with dementia to actively make music.

Further considerations are also necessary with respect to diagnosis, as behavior and psychiatric symptoms are not the same in each form of dementia. For example, there can be increased agitation in persons with Lewy bodies dementia, whereas short-term memory is more impaired in persons with Alzheimer's disease (Caputo et al., 2008). Garrido et al. (2017) caution: "people with different forms of dementia would not only have different therapeutic needs, but also may respond to music differently". (p. 1130).

## Various Types of Music Experiences

## Background Music

The auditory environment impacts both psychological and physiological well-being (Choiniere, 2010), and can affect a person positively or negatively. For instance, research has demonstrated that hospital sound levels are consistently higher than the recommended levels of noise (Choiniere, 2010; Hurtley, 2009) and noise-induced occupational stress can lead to burnout of nurses (Topf & Dilon, 1988). Perez-Cruz et al. (2012) found that patients, caregivers, and providers similarly preferred the presence of background music over ordinary sounds in outpatient and inpatient care areas.

Environmental music, also known less formally as background music, is music "of any kind that is played while some other activity is going on, so that people do not actively attend to it" (Background music, n.d.). Background music is commonly used in business settings and serves two main purposes: arousal and pleasure (North & Hargreaves, 2008). Business world research has stated that background music alters temporal perception over time (Stratton, 1992; North, Hargreaves, & McKendrick, 1999). For example, Gueguen and colleagues (2007) found that customers spent more time at an open-market stall in the presence of pleasant background music in comparison to a nonmusic environment. Similarly, Caldwell and Hibbert (1999) found that music tempo affected customer behaviors in a restaurant, encouraging customers to remain longer in the setting, and thus order more food. These findings suggest the possible affordances of background music in healthcare settings. Specifically, for people living with dementia these affordances include environmental modification to reduce anxiety during care provision or medical procedures; distraction from ordinary facility noises; as enhancement of the dining experience and creation of a relaxed, familiar setting to increase food consumption. These examples are validated by Whear et al.'s (2014) systematic review of studies using relaxing music to soothe residents with dementia during mealtime at long-term care facilities. Results showed a decline in behavioral symptoms such as physically aggressive and nonaggressive behaviors, verbal agitation, and hiding or hoarding behaviors (Whear et al., 2014). This is not a new area of investigation, as a number of studies have concluded that the use of recorded music in long-term care facilities during mealtimes or morning care routines, or at other times of the day, has resulted in diminished levels of agitated behaviors such as wandering, yelling, or other disruptive vocalizations (Bailey & Davidson, 2003; Hillman, 2002; Clark, Lips, & Bilbrey, 1998; Denney, 1997; Goddaer & Abraham, 1994; Hicks-Moore, 2005; Ragneskog, Kihlgren, Karlsson, & Norberg, 1996; Gerdner, 2000; Gerdner & Swanson, 1993; Hicks-Moore & Robinson, 2008; Tabloski, McKinnon-Howe, & Remington, 1995). Although there are significant benefits to background music, these advantages are lost when the music is continually applied and/or applied with little consideration. For example, continuously playing background music in a facility-wide sound system may fade into the background as "white noise," at best reducing its effect and at worst leading to overstimulation and an increase in agitated behaviors.

## Music for Entertainment

Entertainment is defined as "amusement or pleasure that comes from watching a performer" (Merriam Webster, n.d.), and in this regard music for entertainment requires an audience to consciously attend to an event for enjoyment. Music is excellent for creating a sense of community within a shared space. Music can be used in the background to enhance the mood for a party or intensify a film scene, or as the primary focus of entertainment, such as at music concerts or festivals. In healthcare settings amateur and/or professional musicians are often invited into the space to provide small or large-scale concerts for entertainment and recreation purposes. These concerts can provide benefits to all individuals experiencing the performance—not only residents and their family and friends, but also healthcare workers in the facility. Moss, Nolan, and O'Neill (2007) found that live music in the hospital "enhanced the quality of the aesthetic environment of the hospital, with both patients and staff stating that listening to live music helped them to relax, feel happier and more positive" (p. 634). Clements-Cortés (2017) further examined the effectiveness of professional chamber music concerts for cognitively intact and impaired older adults living in long-term care, as well as participating care providers. Results indicated statistically significant outcomes for reduced pain, increased energy, and mood, and four themes emerged: engagement, enjoyment, special moments with others, and connection and meaning (Clements-Cortés, 2017). These conclusions were not only applicable to the participating older adults and care providers, but were also experienced by the performing musicians in the study.

## Recreational Music

Recreational music experiences can be receptive (such as listening) or active interventions (like singing and playing instruments) that are intended solely for enjoyment purposes. Recreational music differs from both music therapy and music medicine in that the emphasis in the latter interventions is to accomplish health-related goals. Although similarities in music presentation may be found within these different disciplinary practices, the facilitator and the goals distinguish the experiences. That being said, there are instances where secondary health-related benefits arise from engagement in recreational music. In dementia care, recreational music can be implemented by a variety of facilitators, including recreation therapists/therapeutic recreationists, recreation aides, and private and family caregivers.

**Music listening.** More than ever with the digital music revolution, people can access music anytime. There are many modalities in which people can listen to and engage with music. Music can be heard on various devices such as CD players, record players, computers, tablets, MP3 players, and smartphones. People may also choose to listen to music through amplification devices, such as a speaker, or through headphones for a more individual music experience. Associated benefits with music listening for people with dementia include decreased agitation (Cohen-Mansfield, 1986; Vink, 2000) and reduced irritability in group music listening experiences (Suzuki et al., 2004).

In 2014 the movie *Alive Inside: The Story of Music and Memory* prompted a rising global awareness of the importance of music on psychosocial and emotional well-being of people with dementia (Rossato-Bennett, 2014). Certain scenes from the film gained viral online attention, portraying "the awakening" of a person with dementia by simply hearing their favorite or familiar music on an iPod. While there are benefits to music listening, this film minimizes the importance of carefully preparing music experiences. It is this author's belief that the use of headphones for people with dementia is not beneficial and should be used with caution. As a person with dementia is already isolated from the world, using headphones can further sustain the isolation. Further, it is not possible to closely monitor a person's responses to music when they are listening to music through individual headphones. Hence, it is recommended that music listening should be experienced between two or more people through speakers. In addition, individuals who are most susceptible to aural sensitivities should be monitored for possible negative reactions to the music (Clements-Cortés, Pearson, & Chang, 2015; Garrido, Chapter 7 in this volume).

**Recreational music making.** Recreational music making is a group activity that involves active music making. It is intended to allow personal expression, to build group support, and to enhance the quality of life (Bittman, Bruhn, Stevens, Westengard, & Umbach, 2003). For people with dementia, playing small percussion instruments (e.g., maraca, drum, tambourine) may be incorporated into group sing-alongs or exercise programs. A rhythm band is another example of a recreational music-making program, where individuals can play instruments along with recorded or live music. Drum circles are a further example of a recreational music-making activity and are often implemented by a trained drum circle facilitator.

**Singing.** There are many advantages to singing with people with dementia individually or in group settings. For example, singing stimulates the release of oxytocin (Kreutz, 2014), which alleviates feelings of stress and anxiety, and develops trust and bonding (Gordon et al., 2008; Gordon, Zagoory-Sharon, Leckman, & Feldman, 2010; Clements-Cortés, 2014b). Older adults often engage in singing experiences with live musical accompaniment, or with the assistance of a recorded sing-along program specifically designed for people with cognitive impairment. Recreational sing-alongs are typically provided by therapeutic recreationists and/or volunteer musicians.

## Music Medicine

Music medicine is a nonpharmacological intervention that involves medical or therapy personnel implementing a passive music listening experience using prerecorded music (Bradt, Dileo, & Shim, 2013). Music selected for music-medicine purposes may be chosen by healthcare professionals or by the client, or it may be selected from a protocol designed by the facilitator, or based on examples found in the literature. These interventions are intended to target physiological and psychological aspects of health, including modification of physiological indicators such as heart rate and stress. Aural stimulation, such as music and sound in nature, can directly affect physiological and psychological responses (Lee & Clements-Cortés, 2014) without the need for a specialized therapist. Music can be prescribed to activate entrainment responses, in which our internal bodily rhythms unconsciously slow down or speed up to match the external rhythms we hear. Nozaradan,

Peretz, and Mouraux (2012) explored the spontaneous building of beat and meter hypothesized to emerge from the selective entrainment of neuronal populations at beat and meter frequencies, finding "that the rhythmic stimuli elicited multiple steady state-evoked potentials (SS-EPs) observed in the EEG spectrum at frequencies corresponding to the rhythmic pattern envelope" (p. 17572). This concept of entrainment allows healthcare professionals to design music interventions that aid in the enhancement of the autonomic bodily functions for optimal health goals, such as relaxation and sleep.

Music-medicine interventions for people with dementia have been less commonly reported; however, research has shown vibroacoustic stimulation, a specific type of music medicine intervention, as highly beneficial (Bartel, Chen, Alain, & Ross, 2017). In a recent study of 18 people diagnosed with Alzheimer's disease, 40Hz rhythmic sensory sound stimulation was applied in six individual sessions lasting 25 minutes each over a 3-week period. In this short time, results suggested that 40Hz stimulation could lead to increased cognition over time (Clements-Cortés, Ahonen, Evan, Freedman, & Bartel, 2016). Cognition was assessed by administering the St. Louis University Mental Status after each session. Results for each session showed there was an effect size of 0.58 (a total performance gain of approximately 12%). Further themes pointed to improved clarity and alertness.

## Community Music Experiences

There are a variety of experiences that fit into the category of community music and some of these overlap with experiences already discussed. Community music experiences are often led by a trained facilitator specialized in the field of community music or music education. These experiences are less commonly implemented with people with dementia; nevertheless, activities such as a drum circle, or an intergenerational music event, have the potential to create a welcoming atmosphere. To further complicate terminology, there is also a music-therapy approach known as Community Music Therapy (CoMT). This differs from community music as it is provided by a qualified music therapist and focuses on addressing therapeutic goals. Clements-Cortés and Pearson (2014) stated that "one major distinction between community music and CoMT is that a trained therapist can frame community music initiatives from a clinical, therapeutic stance that places the emphasis on client-centred growth and wellness" (p. 96).

Community music centralizes its practice through the concept of "musicking." First coined by Christopher Small (1998), "musicking" defines music as an active phenomenon rather than the conventional, limited understanding of music as organized sound. Music in the community music experience creates meaning based on the social relationships that occur during the musicking event. Specific forms of musicking are utilized in CoMT, including health musicking, which is premised on empowering the client in health promotion (Stige, 2002).

> Community Music Therapy as an arena of professional practice is situated health musicking in community, as a planned process of collaboration between client and therapist with a specific focus upon promotion of sociocultural and communal change through a participatory approach where music as ecology of performed relationships is used in non-clinical and inclusive settings. (Stige, 2003, p. 454)

## Music Ensembles

Music ensembles consist of groups like bands and choirs. Because of the group nature of this particular music experience, music ensemble participation can potentially fall within recreational music making or community music experiences. Again, the overall aim or goal of the music ensemble helps to position it accurately. For example, if the aims of the music ensemble are performing at a professional level and striving for performance excellence, the music ensemble will not be characterized as recreational music making, as the primary goal is not one of recreation. On the other hand, community music therapists may use music ensemble or choir modalities where the focus is on therapeutic processes rather than performances or end products.

Older adults with dementia may have the opportunity to participate in a choir or an orchestral or band ensemble. At present, it appears that choirs are more common for people with dementia than instrumental ensembles, as there is little literature discussing instrumental ensembles in this context. In a multiphase research study for people with dementia, participation in choral singing was found to have numerous benefits including building friendship and companionship, reducing anxiety, and instilling happiness and positive feelings (Clements-Cortés, 2013). Statistically significant increases in energy, mood, and happiness, and decreases in pain (Clements-Cortés, 2014b), as well as decreases in anxiety and pain along with increased mood, were reported for both older adults with dementia and their caregivers following 16 weeks of choral singing (Clements-Cortés, 2015). The choirs in these studies are examples of choirs that fall within the scope of Community Music Therapy practice as they were facilitated by a music therapist.

## Music Education and Music Lessons

Music education and music lessons can occur in individual or group learning environments. Music education encompasses the broader spectrum of music lessons, accounting for specialized programs such as music appreciation. For people with dementia, music appreciation programs designed to educate about a particular music genre (e.g., Broadway music) or an artist are commonly offered in long-term care facilities as well as community centers. Music lessons are occasionally offered to people with dementia, and may include learning to play instruments such as piano or guitar with the goal of cognitive stimulation. However, music lessons are more common to those who have previously played a music instrument as a means to reactivate a learned skill or encouraging a person to continue to play an instrument they have played all their lives (Baird & Thompson, 2018).

## Music Therapy

Music therapy aims to accomplish nonmusical and functional goals encompassing areas of health, wellness, and behavior. Clements-Cortés and Bartel (2015) describe music therapy as "the specialized use of music, sound and sound properties to increase physical, cognitive, socioemotional, and spiritual well-being" (p. 13).

The World Federation for Music Therapy defines music therapy as

> the professional use of music and its elements as an intervention in medical, educational, and everyday environments with individuals, groups, families, or communities who seek to optimize their quality of life and improve their physical, social, communicative, emotional, intellectual, and spiritual health and well-being. Research, practice, education, and clinical training

in music therapy are based on professional standards according to cultural, social, and political contexts. (World Federation of Music Therapy, 2011)

Integral to music therapy is the client-therapist relationship, wherein the therapist assesses the client in order to formulate a treatment plan that meets the goals and objectives of the client (Canadian Association of Music Therapists, 2017). Goals may be outlined in a number of domains including cognitive, communication, emotional, physical, social, and spiritual. Here are six examples of music-therapy goals for people with dementia:

- Fostering social interaction
- Decreasing social isolation
- Reducing anxiety
- Reducing aggressive behaviors
- Stimulating reminiscence
- Stimulating cognitive function

Music therapists use both live and recorded music, but live music is more commonly used as it can be adapted in the moment to meet the changing needs of a client. There are many different models of music therapy (not all of which are listed here), including psychodynamic methods such as Guided Imagery and Music (GIM; Bonny, 1978) and Analytic Music Therapy (Priestly, 1975); medical models such as Neurological Music Therapy (NMT; Thaut, 2005); improvisational models including Nordoff-Robbins (Nordoff-Robbins Centre for Music Therapy, 2018) and Aesthetic Music Therapy (AeMT; Lee, 2014); and a variety of other models including Behavioural Music Therapy (Madsen, Cotter, Madsen, 1968), and Music-Centered Music Therapy (Aigen, 2005). There is also the Eclectic approach to music therapy (Arnason, 2002; Brotons, Sabbatella, & Del Moral Marcos, 2017) which incorporates elements from various models.

For people with dementia there is not one single music-therapy approach that works above all others, but there are models which may be more applicable to this population. For example, the technique of therapeutic singing in the NMT approach has been implemented successfully for people with dementia to help maintain verbal responses and abilities, as well as respiratory capacity (Johnson, 2014). Similarly, the Nordoff-Robbins approach has been shown to facilitate both nonverbal and verbal interactions through therapist-improvised music where the therapist responds to the person with dementia and shared feelings (Aigen et al., 2008). Music-therapy interventions can be receptive or active and depend on the requirements of the client and their abilities, goals, and preferences. Receptive music-therapy interventions provide opportunities for relaxation, such as music listening or sensory-based programming. Active interventions include song writing, song lyric discussion, collaborative music making together by performing precomposed music or improvising, and singing.

The client-therapist relationship is the distinguishing factor between music therapy and music medicine. As the collaborative dialogue between the therapist and client instigates therapeutic change, music medicine differs in that its primary focus is on the intrinsic properties of music for healing. The following three examples demonstrate some possible techniques that may be implemented in a clinical music-therapy session with people with dementia.

1. Singing familiar songs can stimulate reminiscence and encourage active singing by the client. Playing small rhythm instruments may further encourage participation in this type of intervention.
2. Writing a song in an individual or group session may inspire clients to translate their internal state into meaningful words. Activities may involve writing original music or a song parody to a precomposed song. Lyrics and song themes can be generated via discussion with the music therapist.
3. The music therapist may choose to engage in music improvisation on the piano or keyboard in order to offer a receptive music experience. This technique can reduce significant levels of anxiety and instill feelings of relaxation in a client.

## Implications for Care Providers

When working with vulnerable populations, such as people with dementia, care providers are encouraged not to make assumptions regarding the possible effects of music interventions. Clements-Cortés, Pearson, and Chang (2015) have created an information brochure in the hopes of educating healthcare workers, family members, friends and volunteers on how to create safe and effective music playlists for people with dementia. In addition to endorsing safe and effective therapeutic resources, Clements-Cortés et al.

(2015) provide helpful instructions for how to manage emotional reactions to the playlists. Listed below are a number of suggestions in which care providers may implement music experiences for people with dementia.

### Music Experiences in the Provision of Daily Living Assistance and Care

1. *Purposefully turn on and turn off music.* Playing music for a continuous period of time can eliminate its potential meaning, which may result in the fade-out phenomenon.
2. *Use music when providing care such as bathing and toileting.* Care providers may play the music at different volumes, or sing in response to the individual requiring care.
3. *Consider playing relaxing or familiar music during mealtimes to promote a calmer environment* (Douglas & Lawrence, 2015).
4. *Consider playing music for sleeping.* Music for sleeping can initiate a slower neurological brain wave that can potentially assist the sleeping patterns of a person in care, such as falling asleep and staying asleep for longer periods of time (National Sleep Foundation, n.d.). Selecting music that is pulsed at 60 beats per minute or less is recommended. "Therasleep" is an example of a paid music resource that is specifically designed for sleeping purposes (Bradstreet, 2005).

### Music Experiences to Promote Psychosocial Opportunities and Foster Engagement

5. *Be aware of and monitor reactions to music, both positive and negative.* Care providers should monitor positive and negative reactions to music. It is important to observe and respond to both reactions, either trying to implement music with positive reactions in other situations, or changing or ceasing music experiences that result in negative reactions.
6. *Create playlists for various times of the day or for specific reasons.* Creating music playlists for various times of the day or for specific reasons can provide opportunities to engage socially with a caregiver when the music experience is shared. For example, a playlist can be created to stimulate reminiscence and help a person connect with their identity. Individualized playlists can help to facilitate relaxation or stimulate entrainment responses.

7. *Use recorded music for group sing-alongs*. Purchase sing-along DVDs or CDs with popular songs to facilitate an active music experience. Caregivers are encouraged to sing along with the person with dementia to create a shared experience.

8. *Use music to promote exercise*. Consider using music to motivate adherence to exercise or to engage or encourage movements such as tapping, clapping, and stomping feet.

9. *Consider drawing or coloring during or after music listening*. This may allow opportunity for creative nonverbal expression and psychosocial awareness to various artistic forms.

## Music Experiences to Stimulate Cognitive Processes

10. *Consider using a portable vibration device*. The VTS 1000-Vibroacustic Therapy Device (Sound Oasis, 2017) has been shown to maintain and promote cognitive stimulation in individuals with dementia (Clements-Cortés et al., 2017). The vibration produced by the VTS-1000 is felt over the back and is heard in the ears.

Specific frequencies that correspond to healthy brain activity can be played and used to stimulate the brain. This brain stimulation results from the cells in the ears and in the body "feeling" the pulses of vibration and responding with each pulse by sending a neural signal to the brain. . . . This stimulation increases the number of brain cells firing at the target rate resulting in a natural support to the desired healthy brain state. (Sound Oasis, 2017)

## Conclusion

When focusing on the potential health benefits of music participation for people living with dementia, it is important to understand the continuum of music experiences available. This continuum spans recreational to therapeutic and passive to active music participation opportunities. With this knowledge caregivers will be better able to select the most appropriate musical experiences for the people with dementia they care for, as well as assess any potential risks or harm that may arise from the inappropriate use of musical experiences by untrained professionals. Various healthcare professionals can provide many different music participation opportunities as described in this chapter. Benefits and considerations for the various experiences from across the continuum have been presented together with implications for care providers and music considerations.

## References

Aigen, K. (2005). *Music centered music therapy.* New Braunfels, TX: Barcelona.

Aigen, K., Miller, C. K., Kim, Y. Psiali, V., Kwak, E. M., & Tague, D. B. (2008). Nordoff-Robbins music therapy. In Alice Ann Darrow (Ed.), *Introduction to approaches in music therapy* (2nd ed., pp. 61–77). Silver Spring, MD: American Music Therapy Association.

Aldridge, D. (1996). *Music therapy research and practice in medicine.* London: Jessica Kingsley.

Arnason, C. (2002). An eclectic approach to the analysis of improvisations in music therapy. *Music Therapy Perspectives, 20,* 4–12.

Bailey, B. A., & Davidson, J. W. (2003). Amateur group singing as a therapeutic instrument. *Nordic Journal of Music Therapy, 12*(1), 18–32. doi:10.1080/08098130309478070

Baird, A., & Thompson, W. F. (2018). The impact of music on the self in dementia. *Journal of Alzheimer's Disease, 61*(3), 827–841.

Background music. (n. d). In *Collins Dictionary Online.* Retrieved from https://www.collinsdictionary.com/dictionary/english/background-music

Bartel, L. R., Chen, R., Alain, C., & Ross, B. (2017). Vibroacoustic stimulation and brain oscillation: From basic research to clinical application. *Music Medicine, 9*(3), 153–166.

Bittman, B., Bruhn, K. T., Stevens, C., Westengard, J., & Umbach, P. O. (2003). Recreational music-making: A cost-effective group interdisciplinary strategy for reducing burnout and improving mood states in long-term workers. *Advances in Mind-Body Medicine, 19*(3–4), 4–15.

Bonny, H. (1978). *Facilitating GIM sessions.* Salina, KS: Bonny Foundation.

Bradstreet, D. (2005). *Therasleep* [CD]. Theramusic.

Bradt, J., Dileo, C., & Shim, M. (2013). Music interventions for preoperative anxiety. *Cochrane Database Systematic Review.* doi:10.1002/14651358.CD006908.pub2

Brotons, M., Sabbatella, P., & Del Moral Marcos, M. T. (2017). Music therapy as a profession in Spain: Past, present and future. *Approaches: Music Therapy & Special Education Music, 9*(1), 111–119.

Bruer, R. A., Spitznagel, E., & Cloninger, C. R (2007). The temporal limits of cognitive change from music therapy in elderly people with dementia or dementia-like cognitive impairment: A randomized controlled trial. *Journal of Music Therapy, 44*(4), 308–328.

Bunt, L., & Hoskyns, S. (2013). *The handbook of music therapy.* New York: Routledge.

Caldwell, C., & Hibbert, S. (1999). Play that one again: The effect of music tempo on consumer behavior in a restaurant. *European Advances in Consumer Research, 4,* 58–62.

Canadian Association of Music Therapists. (2017). About music therapy: Music therapist: clinical process. Retrieved from http://www.musictherapy.ca/about-camt-music-therapy/about-music-therapy/

Caputo, M., Monastero, R., Mariani, E., Santucci, A., Mangialasche, F., Camarda, R., . . . Mecocci, P. (2008). Neuropsychiatric symptoms in 921 elderly subjects with dementia: A comparison between vascular and neurodegenerative. *Acta Psychiatrica Scandinavica, 117*(6), 455–464.

Choiniere, D. B. (2010). The effects of hospital noise. *Nursing Administration Quarterly*, *34*(4), 327–333. doi:10.1097/NAQ.0b013e3181f563db

Cision PR Newswire. (2014, January 23). Setting the record straight: What music therapy is and is not. *PR Newswire*. Retrieved from http://www.prnewswire.com/news-releases/setting-the-record-straight-what-music-therapy-is-and-is-not-241652311.html

Clark, M. E., Lipe, A. W., & Bilbrey, M. (1998). Use of music to decrease aggressive behavior in people with dementia. *Journal of Gerontological Nursing, 24*(7), 10–17.

Clements-Cortés, A. (2013). Buddy's glee club: Singing for life. *Activities, Adaptation & Aging, 37*(4), 273–290. doi:10.1080/01924788.2013.845716

Clements-Cortés, A. (2014a). Sing-a-long DVD and activity package pilot study with older adults. *Music Technology and Education, 7*(2), 123–139.

Clements-Cortés, A. (2014b). Buddy's glee club two: Choral singing benefits for older adults. *Canadian Journal of Music Therapy, 20*(1), 85–109.

Clements-Cortés, A. (2015). Singing for health, connection and care. *Music and Medicine, 7*(4), 13–23.

Clements-Cortés, A. (2017). Artful wellness: Attending chamber music concert reduces pain and increases mood and energy for older adults. *The Arts in Psychotherapy, 52*, 41–49.

Clements-Cortés, A., Ahonen, H., Evan, M., Freedman, M., & Bartel, L. (2016). Short-term effects of rhythmic sensory stimulation in Alzheimer's disease: An exploratory pilot study. *Journal of Alzheimer's Disease, 52*(2), 651–660. doi:10.3233/JAD-160081

Clements-Cortés, A., Ahonen, H., Evans, M., Tang-Wai, D., Freedman, M., & Bartel, L. (2017). Can rhythmic sensory stimulation decrease cognitive decline in Alzheimer's disease? A clinical case study. *Music and Medicine, 9*(3), 174–177.

Clements-Cortés, A., & Bartel, L. (2015). Sound stimulation in patients with Alzheimer's disease. *Annals of Long Term Care, 23*(5), 10–16.

Clements-Cortés, A., & Pearson, S. (2014). Discovering community music therapy in practice: Case reports from two Ontario hospitals. *International Journal of Community Music, 7*(1), 93–111.

Clements-Cortés, A., Pearson, C., & Chang, K. (2015). *Creating effective music listening opportunities*. Toronto, Ontario: Baycrest. Retrieved from http://www.baycrest.org/wp-content/uploads/Creating-Effective-Listening-Opportunities.pdf

Cohen-Mansfield, J. (1986). Agitated behaviors in the elderly. II. Preliminary results in the cognitively deteriorated. *Journal of American Geriatrics Society, 34*(10),722–727.

Denney, A. (1997). Quiet music: An intervention for mealtime agitation? *Journal of Gerontological Nursing, 23*(7), 16–23.

Douglas, J. W., & Lawrence, J. C. (2015). Environmental considerations for improving nutritional status in older adults with dementia: A narrative review. *Journal of the Academy of Nutrition and Dietetics, 115*(11), 1815–1831. https://doi.org/10.1016/j.jand.2015.06.376

Entertainment. (n. d). In *Merriam Webster Dictionary*. Retrieved from http://www.merriam-webster.com/dictionary/entertainment.

Garrido, S., Dunne, L., Chang, E., Perz, J., Stevens, C. J., & Haertschc, M. (2017). The use of music playlists for people with dementia: A critical synthesis. *Journal of Alzheimer's Disease, 60*(3), 1129–1142. doi:10.3233/JAD-170612

Gerdner, L. A. (2000). Effects of individualized versus classical "relaxation" music on the frequency of agitation in elderly people with Alzheimer's disease and related disorders. *International Psychogeriatrics, 12*(1), 49–65. doi:10.1017/S1041610200006190

Gerdner, L. A., & Swanson, E. A. (1993). Effects of individualized music on confused and agitated elderly patients. *Archives of Psychiatric Nursing, 7*(5), 284–291.

Goddaer, J., & Abraham, I. L. (1994). Effects of relaxing music on agitation during meals among nursing home residents with severe cognitive impairment. *Archives of Psychiatric Nursing, 8*(3), 150–158.

Gordon, I., Zagoory-Sharon, O., Leckman, J. F., & Feldman, R. (2010). Oxytocin and the development of parenting in humans. *Biological Psychiatry, 68*(4), 377–382.

Gordon, I., Zagoory-Sharon, O., Schneiderman, I., Leckman, J. F., Weller, A., & Feldman, R. (2008). Oxytocin and cortisol in romantically unattached young adults: Associations with bonding and psychological distress. *Psychophysiology, 45*(3), 349–352.

Groene, R. W. (1993). Effectiveness of music therapy 1:1 intervention with individuals having Senile Dementia of the Alzheimer's type. *Journal of Music Therapy, 30*(3), 138–157.

Gueguen, N., Jacob, C., Lourel, M., & Guellec, H. L. (2007). Effect of background music on consumer's behaviour: A field experiment in an open-air market. *European Journal of Scientific Research, 16*(2), 268–272.

Hicks-Moore, S. L. (2005). Relaxing music at mealtime in nursing homes: Effect on agitated patients with dementia. *Journal of Gerontological Nursing, 31*(12), 26–32.

Hicks-Moore, S. L., & Robinson, B. A. (2008). Favorite music and hand massage: Two interventions to decrease agitation in residents with dementia. *Dementia, 7*(1), 95–108. doi:10.1177/1471301207085369

Hillman, S. (2002). Participatory singing for older people: A perception of benefit. *Health Education Journal, 102*(4), 163–171. doi:10.1108/09654280210434237

Hogan, B. (1998). *Music therapy at the end of life . . . five traditions.* Newbury Park, CA: Sage.

Hurtley, C. (2009). *Night noise guidelines for Europe.* World Health Organization. Retrieved from http://www.euro.who.int/__data/assets/pdf_file/0017/43316/E92845.pdf

Jennings, B., & Vance, D. (2002). The short-term effects of music therapy on different types of agitation in adults with Alzheimer's. *Activities, Adaption, and Aging, 24*(4), 27–33. doi:10.1300/J016v26n04_03

Johnson, S. (2014). Therapeutic singing. In M. Thaut & V. Hoemberg (Eds.), *Handbook of Neurologic Music Therapy* (pp. 140–145). Oxford: Oxford University Press.

Koger, S. M., & Brotons, M. (2000). The impact of music therapy on language functioning in dementia. *Journal of Music Therapy, 37*(3), 183–195.

Kreutz, G. (2014). Does singing facilitate social bonding? *Music and Medicine, 6*(2), 51–60.

Ledger, A. J., & Baker, F. A. (2007). An investigation of long-term effects of group music therapy on agitation levels of people with Alzheimer's disease. *Aging & Mental Health, 11*(3), 330–338. doi:10.1080/13607860600963406

Lee, C., & Clements-Cortés, A. (2014). Applications of clinical improvisation and aesthetic music therapy in medical settings: An analysis of Debussy's "L'isle joyeuse." *Music and Medicine, 6*(2), 61–69.

Lee, C. A. (2014). Aesthetic music therapy. In J. Edwards (Ed.), *The Oxford Handbook of Music Therapy* (pp. 515–537). Oxford: Oxford University Press.

Madsen, C. K., Cotter, V. A., & Madsen, C. H., Jr. (1968). A behavioral approach to music therapy. *Journal of Music Therapy, 5*, 69–71.

McDermott, O., Orrell, M., & Mette Ridder, H. (2014). The importance of music for people with dementia: The perspectives of people with dementia, family carers, staff

and music therapists. *Aging & Mental Health, 18*(6), 706–716. http://dx.doi.org/
10.1080/13607863.2013.875124

Mitchell, L. A., & MacDonald, R. A. (2006). An experimental investigation of the effects
of preferred and relaxing music listening on pain perception. *Journal of Music Therapy,
43*(4), 295–316.

Moss, H., Nolan, E., & O'Neill, D. (2007). A cure for the soul? The benefits of live music in
the general hospital. *Irish Medical Journal, 100*(10), 634–636.

National Sleep Foundation. (n.d.). *Can music help you calm down and
sleep better?* Retrieved from https://sleepfoundation.org/sleep-topics/
can-music-help-you-calm-down-and-sleep-better

Nordoff-Robbins Centre for Music Therapy. (2018). *Nordoff-Robbins music therapy.*
Retrieved from http://steinhardt.nyu.edu/music/nordoff/therapy/nordoff

North, A. C., & Hargreaves, D. J. (2008). *The social and applied psychology of music.*
New York: Oxford University Press.

North, A., Hargreaves, D., & McKendrick, J. (1999). Music and on-hold waiting time.
*British Journal of Psychology, 90*, 161–164.

Perez-Cruz, P., Nguyen, L., Rhodali, W., Hui, D., Palmer, J., Sevy, I., . . . Bruera, E. (2012).
Attitudes and perceptions of patients, caregivers, and health care providers toward
background music in patient care areas: An exploratory study. *Journal of Palliative
Medicine, 15*(1), 1130–1136.

Priestley, M. (1975). *Music therapy in action.* London: Constable.

Raglio, A. (2010). Music therapy in dementia. In Dementia: *Non-pharmacological therapies
in dementia, 1*(1), Nova Publications. Retrieved from: https://pdfs.semanticscholar.org/
10cb/d19b64a8a670bf0191835b46fbea2eee38fb.pdf

Raglio, A., Bellelli, G., Traficante, D., Gianotti, M., Ubezio, M. C., Villani, D., & Trabucchi,
M. (2008). Efficacy of music therapy in the treatment of behavioral and psychiatric
symptoms of dementia. *Alzheimer Disease and Associated Disorders, 22*(2), 158–162.
doi:10.1097/WAD.0b013e3181630b6f

Ragneskog, H., Kihlgren, M., Karlsson, I., & Norberg, A. (1996). Dinner music for demented patients: Analysis of video-recorded observations. *Clinical Nursing Research,
5*(3), 262–282.

Rossato-Bennett, M. (Director). (2014). *Alive inside: The story of music and memory*
[Motion picture on DVD]. USA: Projector Media.

Sixsmith, A., & Gibson, G. (2007). Music and the well-being of people with dementia.
*Ageing and Society, 27*(1), 127–145. doi:10.1017/S0144686x06005228

Small, C. (1998). *Musicking: The meanings of performing and listening.* Hanover,
NH: University Press of New England.

Sound Oasis. Vibroacoustic therapy systems. Retrieved from http://www.soundoasis.com/
product40category/vibroacoustic-therapy-systems/

Stige, B. (2002). *Culture-centered music therapy.* Gilsum, NH: Barcelona.

Stige, B. (2003). *Elaborations toward a notion of community music therapy.* Oslo,
Norway: Unipub.

Stratton, V. N. (1992). Influence of music and socializing on perceived stress while
waiting. *Perceptual and Motor Skills, 75*(1), 334.

Suzuki, M., Kanamori, M., Watanabe, M., Nagasawa, S., Kojima, E., Ooshiro, H., &
Nakahara, D. (2004). Behavioral and endocrinological evaluation of music therapy for
elderly patients with dementia. *Nursing & Health Sciences, 6*(1), 11–18.

Tabloski, P. A., McKinnon-Howe, L., & Remington, R. (1995). Effects of calming music on the level of agitation in cognitively impaired nursing home residents. *American Journal of Alzheimer's Disease and Other Dementias, 10*(1), 10–15.

Thaut, M. H. (2005). *Rhythm, music and the brain.* New York and London: Taylor and Francis.

Topf, M., & Dillon, E. (1988). Noise-induced stress as a predictor of burnout in critical care nurses. *Heart & Lung: The Journal of Critical Care, 17*(5), 567–574.

Van de Winckel, A., Feys, H., De Weerdt, W., & Dom, R. (2004). Cognitive and behavioural effects of music-based exercises in patients with dementia. *Clinical Rehabilitation, 18*(3), 253–260.

Vink, A. (2000). The problem of agitation in elderly people and the potential benefit of music therapy. In D. Aldridge (Ed.), *Music therapy in dementia care* (2nd ed., pp. 102–118). London, UK: Jessica Kingsley.

Whear, R., Abbott, R., Thompson-Coon, J., Bethel, A., Rogers, M., Hemsley, A., . . . Stein, K. (2014). Effectiveness of mealtime interventions on behavior symptoms of people with dementia living in care homes: A systematic review. *Journal of the American Medical Directors Association, 15*(3), 185–193

World Federation of Music Therapy. (2015). *What is music therapy?* Retrieved from http://www.wfmt.info/wfmt-new-home/about-wfmt/

Yinger, O. S., & Gooding, L. F. (2015). A systematic review of music-based interventions for procedural support. *Journal of Music Therapy, 52*(1), 1–77.

# 2

# Is Music Special for People With Dementia?

*Matthieu Ghilain, Loris Schiaratura, Ashmita Singh,*
*Micheline Lesaffre, and Séverine Samson*

## Introduction

A growing number of aged care facilities offer musical interventions for people with dementia in order to soothe them and improve well-being. Not only are these interventions pleasure inducing for the patients, even in the severe stage (Zentner, Grandjean, & Scherer, 2008), but they also appear to have a positive impact on the distress of caregivers (Narme et al., 2014; Särkämö et al., 2014, 2016). Altogether, the field of musical interventions in dementia care has yielded a lot of valuable knowledge in several domains such as behavior, emotion, and cognition (for further discussion, see also Särkämö's and Lipe & Edmonston's chapters in this book).

Nonetheless, as noted in a recent meta-analysis (van der Steen et al., 2017), the scientific validity of present-day literature is often a point of debate, and indeed results should be interpreted prudently. The size of the effects remains modest and the potency of these treatments vary considerably between individuals, the reason for which is currently unknown. Moreover, the magnitude of the effects is often exacerbated by numerous methodological biases, such as the small number of participants, the lack of blind assessment, and the absence of nonmusic or another intervention to control for change due to patients' stimulation (Samson, Clément, Narme, Schiaratura, & Ehrlé, 2015). Other factors related to the specific characteristics of music, such as the style and familiarity of the musical excerpts, tempo, and meter, could also influence results, but at present little is known about the impact of these variables on the effectiveness of musical support. The causes behind different effects observed after receptive and participatory musical interventions are also currently difficult to pinpoint. Overall, many variables manipulated during investigations are not bound by universal standards, making generalizations across studies, as well as populations, difficult, if not impossible. Therefore, further research is necessary: not simply focusing on musical interventions but also towards answering basic science questions of how humans internalize, experience, and produce music. This will enhance knowledge about the mechanisms behind observed effects, since these are not yet understood. One such foundational phenomenon is the ability to progressively move in synchrony with an environmental system. Known as rhythmic entrainment, this mechanism is very relevant to the topic of musical interventions, particularly for dementia.

In this chapter, we will focus on rhythmic entrainment, including synchronization to music. An overview on the impact of musical rhythms on various domains (motor, cognition, emotion, and social functioning) will be proposed and followed by a presentation of the effect of music listening on the motor responses of people living with dementia. Studies that have investigated how social interaction modulates the rhythmic entrainment will also be discussed. Ultimately, some recent perspectives for assessing the therapeutic benefits of music will be examined.

## Rhythmic Entrainment

Rhythmic entrainment is defined as the progressive change towards synchronization between two physical or biological systems that interact with each other. It includes the perfect synchronization of period (cycles of the two systems having the same duration) and must be in-phase (rhythmic events of the same period occurring at the same time). Rhythmic entrainment is the process of reaching this state of synchronization due to the changes in two dynamical systems as they interact with each other, rather than the simple existence of phase locking and period synchronization between two things (Clayton, Sager, & Will, 2005; Leman, Buhmann, & Van Dyck, 2017; Trost, Labbé, & Grandjean, 2017). Music listening frequently encourages people to move or clap their hands (Drake, Penel, & Bigand, 2000). These responses to the rhythm of an environmental stimulus illustrate the phenomenon of rhythmic entrainment and have also been observed among people with dementia (Holmes, Knights, Dean, Hodkinson, & Hopkins, 2006; Lesaffre, Moens, & Desmet, 2017; Sherratt, Thornton, & Hatton, 2004). Several lines of evidence suggest that rhythmic entrainment induced by listening to music seems to modulate not only the motor system, but also cognitive, emotional, autonomic, and social functioning.

The influence of musical rhythms on motor responses and in particular on sensorimotor synchronization (SMS) is well known (Leman et al., 2017; Repp, 2005; Repp & Su, 2013; Zatorre, Chen, & Penhune, 2007). SMS to music corresponds to the coordination of movement at the occurrence of the beat of music (Repp, 2005). In this context, the perception of beats organized according to a well-established hierarchy, notably by metrics induced by music, prepares the motor system to respond (Grahn & Brett, 2007). Several brain areas are involved in this process: the basal ganglia play a role in the processing of rhythm and motor control (Grahn, 2009; Grahn & Rowe, 2009; Nombela, Hughes, Owen, & Grahn, 2013), while the cerebellum and supplementary motor area participate in coordinating to musical rhythm (Chen, Penhune, & Zatorre, 2008). According to Jones' Dynamics Attending Theory (1987), when we listen to music, we detect the organization of music beats through the repetition of metric structure, perceptual accents, and rhythm integer ratios. The perception of such repetitions generates neural synchronies relating to temporal expectations between the auditory and motor systems (Coull, Cheng, & Meck, 2011; Nozaradan, Peretz, Missal, & Mouraux, 2011). These neural synchronies in the primary sensorimotor, pre-motor, and supplementary motor areas lead to audio-motor coupling (Chen et al., 2008), which is required for the perception and integration of sensorimotor information necessary for the production of coordinated movements in time with the musical beat.

The motor and sensorimotor regions involved in synchronization to music also seem to play a role in cognition. Interactions between the supplementary motor area and the putamen are related to both motor performance and executive functions (Marchand et al., 2013). Moreover, the cerebellum, involved in precision coordination, is also involved in processes related to executive and attentional functions (Baillieux, Smet, Paquier, De Deyn, & Mariën, 2008). SMS, by stimulating the aforementioned brain networks, could inadvertently also modulate cognitive functioning. Therefore, questions about possible links between music synchronization and cognitive functioning in dementia appear to be warranted.

Other studies have demonstrated that the rhythm of music influences emotions. Music played at a fast tempo is often perceived as more cheerful and stimulating than music played at a slow tempo (Husain, Thompson, & Schellenberg, 2002; Khalfa, Roy, Rainville, Dalla Bella, & Peretz, 2008).

Listening to music also provides pleasure, as many studies have shown (Blood & Zatorre, 2001; Salimpoor, Benovoy, Larcher, Dagher, & Zatorre, 2011; Salimpoor, Zald, Zatorre, Dagher, & McIntosh, 2015; Zatorre, 2015). This same sensation of pleasure can be found in dance (Bernardi, Bellemare-Pepin, & Peretz, 2017), which requires SMS based on the coupling between the auditory and motor systems (Janata, Tomic, & Haberman, 2012). The stimulation of the vestibular system and the connections it maintains with the limbic system could mediate emotional responses (Trost et al., 2017).

Another hypothesis, not exclusive to the previous one, to explain the impact of musical rhythm on emotions is based on the concept of temporal expectations induced by musical listening. Musical structure including but not limited to beat, meter, and rhythm, makes it possible to predict the appearance of events. These anticipatory mechanisms may also trigger neuro-physiological events associated with the feeling of pleasure. As neuroimaging studies have shown, the dorsal striatum may be activated in this phase of pleasure anticipation, accompanied by the release of dopamine, while the ventral striatum is activated during the musical pleasure peak experience (Salimpoor et al., 2011). At this peak, the release of dopamine occurs in the ventral striatum. These physiological responses in the striatum participate in the fronto-striatal network which appears to modulate emotional state and motivation, as evidenced by a recent study of brain imaging coupled with transcranial magnetic stimulation (Mas-Herrero, Dagher, & Zatorre, 2018). Given these results, we suggest that there is a link between the rhythmic entrainment and emotional process that could explain, at least in part, the impact of musical interventions on the emotional state of people living with dementia.

Music can also modify the actions of the autonomic system. Musical listening modulates autonomic rhythms, such as respiratory cycles (Khalfa et al., 2008) and heart rate (Hodges, 2011) and synchronizes them to musical rhythms (Bernardi et al., 2009). Codrons, Bernardi, Vandoni, and Bernardi (2014) observed spontaneous synchronization of respiration to the beat during group musical activities with healthy adults. Their results show that the degree of synchronization coherence is higher during musical sequences than in metronomic sequences, indicating that music has a greater influence on the autonomic system. Khalfa and colleagues (2008) have demonstrated that, in young adults, the changes in skin conductance responses, facial muscle activity, and blood pressure increase more when listening to cheerful music (fast tempo and major mode) than when listening to sad music (slow tempo and minor mode). Overall, there is enough evidence to conclude that music can influence certain autonomic rhythms and that the intensity of the influence may be modulated by the specific properties of the auditory stimulus (like tempo, mode, and whether it is music or metronome cues). To inform current interventions for people with dementia, further investigation of the effects of rhythmic entrainment to endogenous rhythms (and the psycho-physiological responses that follow), particularly in relation to the properties of musical sequences, is needed.

Finally, rhythmic entrainment can also modulate social functioning. It has been shown that interpersonal coordination produced by collectively listening to music can encourage the emergence of social cooperation (Wiltermuth & Heath, 2009). Similarly, synchronizing movements to music during dance can promote empathy (Koehne, Behrends, Fairhurst, & Dziobek, 2016), at least in healthy subjects. By causing individuals to coordinate together to the rhythm of music, it facilitates communication and the expression of social behaviors (Cross & Morley, 2008). This effect may be particularly beneficial for people with dementia as suggested by Cason, Schiaratura, and Samson (2017). Since language disorders are frequently observed in many types of dementia (Cummings et al., 1985) and can lead to communication difficulties, exacerbated by sensory decline associated with aging (Kamil & Lin, 2015), social isolation increases and significantly reduces the quality of life of people with a diagnosis (Klimova, Maresova, Valis, Hort, & Kuca, 2015). By stimulating the coordination of movements and interpersonal entrainment, musical interventions could promote social cohesion and nonverbal expression for people with dementia and facilitate communication, especially when these interventions are conducted collectively. This idea found support in a randomized controlled study carried out in early dementia (Särkämö et al., 2014). The reduction of caregivers' emotional burden reported after regular music-based intervention (singing or music listening during everyday care), was even more evident when the caregiver participated in musical activities with the patient. By strengthening social interactions between patient and caregiver, the benefits of these musical interventions on caregiver distress become more prominent (for further discussion, see also Tamplin and Clark's chapter in this book). Such effects resulting from interpersonal coordination as well as rhythmic entrainments induced by music might be particularly relevant to improve the social relationships between the patients and their medical or family caregivers.

As mentioned above, many studies suggest that rhythmic entrainment, by promoting interactions between an individual and their musical environment, will impact the motor, cognitive, emotional, autonomic, and social systems. Further investigation of possible links between rhythmic entrainment during musical interventions and these different systems will be relevant in planning activities and aid in better understanding the therapeutic effects of current musical interventions in dementia care.

## Effect of Music on Motor Responses in People With Dementia

Few studies provide solid arguments in favor of rhythmic entrainment during musical interventions for people with dementia. Nevertheless, it has been reported numerous times that even in advanced stages of the disease, people move to the rhythm of music spontaneously (Holmes et al., 2006; Lesaffre et al., 2017; Sherratt et al., 2004) or intentionally (Ghilain et al., in preparation). This suggests that the musical environment facilitates the rhythmic entrainment of people with dementia. This act of rhythmically entraining and in particular SMS to music might elicit pleasure, but this is yet to be confirmed in this population. The social environment linked to the presence of other people (musicians or not) could also facilitate this rhythmic entrainment. Among the studies reported in literature, the influence of social context and interpersonal coordination on motor responses has attracted the attention of researchers. Few authors have attempted to answer this question in relation to dementia specifically by measuring the frequency and duration of motor responses and their engagement to music (Sherratt et al., 2004) or the overall body movements (Holmes et al., 2006; Lesaffre et al., 2017), along with hand movements (Ghilain et al., in preparation) in response to music. The design and results of these experiments will be discussed in detail in the sections to come.

## Frequency and Duration of Motor Responses and Engagement to Music

In an initial study, Sherratt and collaborators (2004) compared the spontaneous responses (including rhythmic movements) of people with dementia (n = 24) to familiar songs under four listening conditions: the first condition was performed in the presence of a musician singing in front of the patients; the second was a prerecorded auditory version of singing from the musician who was not physically present; the third was the original (commercial) auditory recording; and the last was a silent condition without music or singer. The task took place in a common space where participants were free to interact with others (patients, caregivers). Two observers independently identified specific responses to music (singing, whistling/humming, and movements in response to music, such as clapping) and other actions not related to music (nonrhythmic movements); well-being behaviors (assessed in five categories from extreme ill-being to extreme well-being); interaction with others; and passive (e.g., sleeping) or symptomatic behaviors (e.g., wandering, perseveration, individually defined behaviors like moving furniture), based on a pre-established decoding method (Kitwood, 1997). These different behaviors, which can be associated with nonverbal communication indices, were analyzed in terms of duration (percentage of the time the behavior took place in relation to the total duration of observation) and frequency (number of repetitions of the behavior during the total duration period). According to the results, the duration of the movements produced in response to the music, as well as the duration of the well-being ratings of people with dementia, were significantly higher in the singer's presence compared with the audio recording (no musician present) and the silent condition. On the other hand, interactions with other people were more frequent in the singer's absence (auditory recordings), suggesting that people with dementia seek to share their musical experience and socialize with other individuals when the singer is not present.

In a similar study, Holmes and colleagues (2006) also used a behavioral index coding method to compare the quality of engagement in people with dementia (n = 32) under the same conditions (a musician-present condition, a prerecorded condition of the same musician, and a silent condition) as above. Engagement measurement was rated on a six-point Likert scale ranging from a high level of engagement and self-expression (associated with high well-being), to apathy, rage, or despair (associated with extreme ill-being). The results showed that the percentage of people with dementia showing positive engagement was higher in the presence of a musician than in the silent condition. However, the ratings during the recorded condition did not differ from the silent condition. Interestingly, positive engagement during musical interventions was higher in the presence of musicians than in the recorded condition, but this was only evident in people with severe dementia. These results align with the study by Sherratt and colleagues (2004) and highlight the importance of social interaction in spontaneous motor responses to music by people living with dementia.

## Overall Body Movement

Lesaffre and collaborators (2017) examined spontaneous movements (including synchronization movements) in response to music by people with dementia (n = 32). As in the two previous studies, the authors' objective was to verify the impact of a singer's presence on the participants' behavior, but their dependent measure was different. In this case, the authors used a special experimental device developed at the University of Ghent called the "Music Balance Board." Composed of a chair placed on a board equipped with motion sensors, this device offers the possibility of recording the movements produced naturally in response to the music. It enables the calculation of an overall motor score (Quantity of Motion, QOM) estimated from the variations in weight exerted on the board during the person's movements. Four participants with dementia were recorded simultaneously while listening to popular songs and excerpts of classical music performed by a musician (singing in front of them) or through a prerecorded audio version sung by the same musician (who was not physically present). The results showed that the patients' movement score (QOM) was higher in the presence of the musician than in their absence. The impact of the social context complements the results previously reported (Holmes et al., 2006; Sherratt et al., 2004).

It is imperative to acknowledge that the experimental conditions (live and prerecorded) of these studies are not perfectly comparable. They differ not only in the presence or absence of a musician, but also in the number of sensory modalities engaged. The presence of the musician is a multimodal (audio-visual) condition, whereas the prerecorded condition is limited in this case to a unimodal (auditory) condition. Therefore, on the basis of this data, it is not possible to confirm the influence of social context on spontaneous movements and expressiveness of people with dementia in response to music. Ultimately, these different measures provide information on the production and intensity of movements in response to the music without indicating whether these motor responses are synchronized or not to the rhythm of the music. It is therefore necessary to investigate this line of

research while maintaining a context of entrainment by utilizing more precise tools to measure sensorimotor activity.

## Accuracy and Consistency of Hand
## Movement Synchronization

The sensorimotor synchronization (SMS) response to music can be measured by asking participants to tap to the beat of the rhythm on a tablet with
their hand or finger. This enables rigorous measurement of synchronization in terms of precision and consistency (the opposite of variability) of the
finger/hand tapping. This measure provides an indicator of an individual's
rhythmic entrainment with the environment. To date, most research has
compared SMS to metronome cues and music in healthy adults (Repp &
Su, 2013) and people with Parkinson's disease (Bégel, Verga, Benoit, Kotz,
& Dalla Bella, 2018; Benoit et al., 2014; Dalla Bella, Dotov, Bardy, & Cochen
de Cock, 2018; Merchant, Luciana, Hooper, Majestic, & Tuite, 2008). To the
best of our knowledge, there are no published data regarding SMS to music
among people living with dementia.

Continuing the work of Lesaffre and collaborators (2017), we also used
a Music Balance Board that has been improved to measure sensorimotor
synchronization through an additional motion sensor and the use of a
video camera (Figure 2.1). A tablet placed on the armrest of the chair and
connected to a motion sensor makes it possible to record the participant's
tapping to music or metronome cues (as described by Desmet, Lesaffre, Six,
Ehrlé, & Samson, 2017). Using a large screen to project a life-size audio-
visual recording of a musical performance (for the prerecorded condition)
allows for the comparison of SMS in the presence or absence of a musician.

The current study, carried out in Belgium (Woon- en Zorgcentrum Home
Sint-Franciscus in Kluisbergen) with 26 participants with a diagnosis of dementia (Alzheimer's disease, vascular or mixed dementia), has assessed
the feasibility of this tool (Ghilain et al., in preparation). Fine synchronization measurements (accuracy and consistency) of responses were collected
under four experimental conditions where patients were asked to synchronize hand movements with rhythmic sequences (metronomic or musical) in
front of a singer who performed the task with them, or with the audio-visual
recordings (AV) without the presence of the singer. The results showed that
SMS to metronome cues is more accurate and more consistent (less variable)

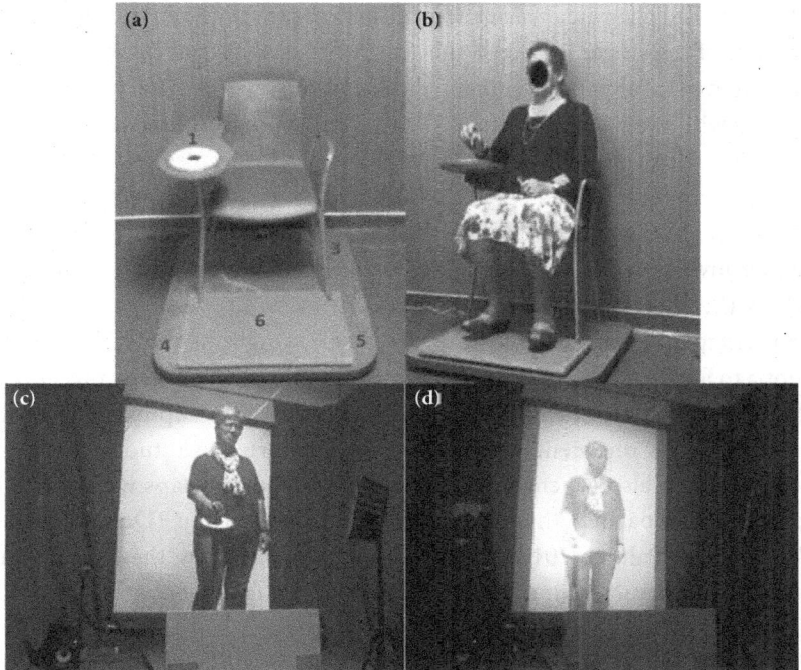

**Figure 2.1.** Illustration of the experimental setup. (**a**) depicts the different sensors: sensor 1 detects the taps of the hand and the overall body movements are caught by the other sensors (from 2 to 6). In (**b**) the participant taps and moves naturally in response to rhythmic sequences. In (**c**) a live partner taps and moves naturally in response to rhythmic sequences. (**d**) shows a screen with an audio-visual recording of the partner.

than SMS to music, which is in agreement with data obtained in Parkinson's disease patients (Bégel et al., 2018, Benoit et al., 2014). As music presents a less salient metric than that of a metronome (Rodger & Craig, 2016), the difference between SMS to metronome cues and to music performance can probably be explained by the greater facility to anticipate the beat in a metronome rather than in a musical sequence.

In addition, the analyses revealed an interaction between the rhythmic sequence environment and the social environment on SMS. SMS to metronome cues was more accurate and more consistent during the AV recording than when the singer was present, which was not the case for SMS to music, SMS to music being more accurate and more consistent in the singer's actual presence than with the AV recording. Therefore, it seems that the influence of the social environment varies according to the rhythmic environment. The presence of the singer improves the accuracy and consistency of synchronization to music for people with dementia, but the reverse pattern is observed for SMS with the metronome. SMS with complex rhythmic sequences, such as music, are more attention-intensive than SMS to simpler sequences, such as the metronome (Chapin et al., 2010; Chen et al., 2008). Having more difficulty synchronizing to the music than to metronome, the presence of a live partner involving joint attention (Sebanz, Bekkering, & Knoblich, 2006) as well as the sharing of the coordination intention between the singer and the participant (Tomasello & Carpenter, 2007) might help people with dementia to focus their attention on this more demanding task. The presence of a live partner instead of an AV recording can therefore improve SMS to the music rhythm. Consistent findings have been reported in a study with young musicians that also synchronized more accurately their taps with music in front of a live partner than in front of recorded partner (Demos, Carter, Wanderley, & Palmer, 2017). As suggested by the authors, the bidirectional interactions characterizing the live condition may facilitate synchronization to the music.

Similarly, in our present study, we can suggest that the bidirectional interactions between the singer and the participant, which are not available in the prerecorded condition, may optimize SMS to music by providing additional cues to tap in synchrony with musical rhythm. SMS to metronome, which is simpler but also less attractive than SMS to music, is not improved by the live partner, which decreases the SMS performance. With isochronous and unattractive sequences, people with dementia seem to be distracted rather than stimulated by the social interaction. Taken together, the results of this study demonstrate that rigorous measures of SMS can be recorded in people with dementia yielding a method to study the rhythmic entrainment of individuals with various environments in the clinical population.

According to Leman and collaborators (2017), the expressiveness and motivation to move to the rhythm appears to be unique to music. As previously emphasized, music induces a joint reaction that shows a certain level of nonverbal communication between individuals. The mutual interaction of individuals adds to the driving force of music towards stimulating synchronization. For these reasons, music may facilitate more social interaction and interpersonal coordination than the metronome. Nevertheless, this difference is yet to be demonstrated by formal scientific investigation. Future studies would need to complement SMS measurements with the analysis of nonverbal communication indices of people with dementia.

## Conclusion and Perspectives

In a time where pharmacological options for the management of behavioral, psychological, and social symptoms of dementia are expensive, moderately effective, and often accompanied by an array of side effects, nonpharmacological interventions offer a low-risk, easily accessible solution. Many different types of social and pleasure-inducing interventions are available today for people living with dementia, such as cooking, visual art, music, and so forth. As previously explained in this chapter, rhythmic entrainment to music, in the form of SMS, modulates motor responses, cognition pathways, and autonomic functioning; it induces pleasure through mediation of emotional responses and temporal anticipatory mechanisms; and it positively influences interpersonal coordination and social cooperation. Although no studies have directly compared the efficacy of other interventions with music-based interventions that specifically focus on role of entrainment in this population, music may be unique for people with dementia, and reports on therapeutic effects of music in dementia care are already plentiful in literature. As we have seen, music appears to be extensive and quite holistic in its effects and benefits. In this chapter, we argue that rhythmic entrainment with the musical environment and in particular SMS induced by musical activity could be partly responsible for these effects (Trost et al., 2017).

There are few tools currently available to help confirm this hypothesis in the field of dementia. Various complementary methods have been proposed to evaluate rhythm entrainment in people with dementia. In a multi-modal approach, it would be possible to study the relationship that might exist between the different aspects of motor responses. The interest in this method is justified, since it provides the opportunity to form scientifically valid connections between such SMS measures on one side and emotion, cognition, behavior, or social interactions on the other side. Such methodological developments could also provide researchers and clinicians with additional tools to examine and evaluate the effectiveness of rhythmic entrainment (e.g., SMS) in musical interventions.

Information currently available in the field can already be utilized to better adapt musical interventions for people with dementia, or at the very least be used to guide future investigations. Given the influence of social context on SMS and nonverbal communication responses, collective musical activities will likely have more impact than individual activities in dementia care. Moreover, the involvement of caregivers and activities encouraging patient-caregiver bonding can also be incorporated. A possible way of accomplishing this can be to teach caregivers to provide supplemental home-based music sessions. The development of personalized musical interventions for people living with dementia could be useful, but in order to actualize this, more needs to be known about the specific effects that musical properties and social context have in dementia. Ultimately, there is no doubt that the future holds exciting discoveries and insights to be made in this domain, especially as novel tools and innovative solutions start to be developed and utilized across the world.

## Acknowledgments

Grant support was provided by the Regional Council of Hauts-de-France and University Charles de Gaulle Lille III to M.G., the Globalink Research Award from Mitacs Canada to A.S., the Hubert Curien partnership from the Ministry of Foreign Affairs to S.S. and M.L., and the Institut Universitaire de France to S.S. The authors wish to thank Ivan Schepers of Ghent University for development of the hardware. We are especially grateful to our performer Linda Vanderstichele, the personnel of the WZC St. Franciscus at Kluisbergen (B), and the patients who volunteered to participate in the study.

## References

Baillieux, H., Smet, H. J. D., Paquier, P. F., De Deyn, P. P., & Mariën, P. (2008). Cerebellar neurocognition: Insights into the bottom of the brain. *Clinical Neurology and Neurosurgery, 110*(8), 763–773. https://doi.org/10.1016/j.clineuro.2008.05.013

Bégel, V., Verga, L., Benoit, C. E., Kotz, S. A., & Dalla Bella, S. (2018). Test-retest reliability of the Battery for the Assessment of Auditory Sensorimotor and Timing Abilities (BAASTA). *Annals of physical and rehabilitation medicine.* https://doi.org/10.1016/j.rehab.2018.04.001.

Benoit, C. E., Dalla Bella, S., Farrugia, N., Obrig, H., Mainka, S., & Kotz, S. A. (2014). Musically cued gait-training improves both perceptual and motor timing in Parkinson's disease. *Frontiers in human neuroscience, 8,* 494.

Bernardi, L., Porta, C., Casucci, G., Balsamo, R., Bernardi, N. F., Fogari, R., & Sleight, P. (2009). Dynamic interactions between musical, cardiovascular, and cerebral rhythms in humans. *Circulation, 119*(25), 3171–3180. https://doi.org/10.1161/CIRCULATIONAHA.108.806174

Bernardi, N. F., Bellemare-Pepin, A., & Peretz, I. (2017). Enhancement of pleasure during spontaneous dance. *Frontiers in Human Neuroscience, 11.* https://doi.org/10.3389/fnhum.2017.00572

Blood, A. J., & Zatorre, R. J. (2001). Intensely pleasurable responses to music correlate with activity in brain regions implicated in reward and emotion. *Proceedings of the National Academy of Sciences of the United States of America, 98*(20), 11818–11823. https://doi.org/10.1073/pnas.191355898

Cason, N., Schiaratura, L., & Samson, S. (2017). Synchronization to music as a tool for enhancing non-verbal communication in people with neurological diseases. In M. Lesaffre, P.-J. Maes, M. Leman (Eds.), *The Roudledge Companion to Embodied Music Interaction* (pp. 304–312). New York: Routledge. https://doi.org/10.4324/9781315621364.ch33

Chapin, H. L., Zanto, T., Jantzen, K. J., Kelso, S., Steinberg, F., & Large, E. W. (2010). Neural responses to complex auditory rhythms: The role of attending. *Frontiers in Psychology, 1.* https://doi.org/10.3389/fpsyg.2010.00224

Chen, J. L., Penhune, V. B., & Zatorre, R. J. (2008). Moving on time: Brain network for auditory-motor synchronization is modulated by rhythm complexity and musical training. *Journal of Cognitive Neuroscience, 20*(2), 226–239. https://doi.org/10.1162/jocn.2008.20018

Clayton, M., Sager, R., & Will, U. (2005). In time with the music: The concept of entrainment and its significance for ethnomusicology. *European meetings in Ethnomusicology, 11*(1), pp. 3–142.

Codrons, E., Bernardi, N. F., Vandoni, M., & Bernardi, L. (2014). Spontaneous group synchronization of movements and respiratory rhythms. *PLOS ONE, 9*(9), e107538. https://doi.org/10.1371/journal.pone.0107538

Coull, J. T., Cheng, R. K., & Meck, W. H. (2011). Neuroanatomical and neurochemical substrates of timing. *Neuropsychopharmacology, 36*(1), 3.

Cross, I., & Morley, I. R. M. (2008). The evolution of music: Theories, definitions and the nature of the evidence. In S. Malloch & C. Trevarthen (Eds.), *Communicative Musicality* (pp. 61–81). New York: Oxford University Press.

Cummings, J. L., Benson, F., Hill, M. A., & Read, S. (1985). Aphasia in dementia of the Alzheimer type. *Neurology, 35*(3), 394–397.

Dalla Bella, S. D., Dotov, D., Bardy, B., & Cochen de Cock, V. C. (2018). Individualization of music-based rhythmic auditory cueing in Parkinson's disease. *Annals of the New York Academy of Sciences.* https://doi.org/10.1111/nyas.13859

Demos, A. P., Carter, D. J., Wanderley, M. M., & Palmer, C. (2017). The unresponsive partner: Roles of social status, auditory feedback, and animacy in coordination of joint music performance. *Frontiers in psychology, 8.* 149.

Desmet, F., Lesaffre, M., Six, J., Ehrlé, N., & Samson, S. (2017). Multimodal analysis of synchronization data from patients with dementia. In *Proceedings of the ESCOM 2017 conference.* Retrieved from http://hdl.handle.net/1854/LU-8521738

Drake, C., Penel, A., & Bigand, E. (2000). Tapping in time with mechanically and expressively performed music. *Music Perception: An Interdisciplinary Journal, 18*(1), 1–23.

Ghilain, M., Schiaratura, L., Lesaffre, M., Six, J., Desmet, F., & Samson, S. (in preparation). *Effect of social contexts on synchronization to musical rhythms in people with dementia.*

Grahn, J. A. (2009). The role of the basal ganglia in beat perception. *Annals of the New York Academy of Sciences, 1169*(1), 35–45. https://doi.org/10.1111/j.1749-6632.2009.04553.x

Grahn, J. A., & Brett, M. (2007). Rhythm and beat perception in motor areas of the brain. *Journal of Cognitive Neuroscience, 19*(5), 893–906. https://doi.org/10.1162/jocn.2007.19.5.893

Grahn, J. A., & Rowe, J. B. (2009). Feeling the beat: Premotor and striatal interactions in musicians and nonmusicians during beat perception. *Journal of Neuroscience, 29*(23), 7540–7548. https://doi.org/10.1523/JNEUROSCI.2018-08.2009

Hodges, D. A. (2011). Psychophysiological measures. In P. N. Juslin & J. Sloboda (Eds.), *Handbook of music and emotion: Theory, research, applications* (pp. 279–312). New York: Oxford University Press.

Holmes, C., Knights, A., Dean, C., Hodkinson, S., & Hopkins, V. (2006). Keep music live: Music and the alleviation of apathy in dementia subjects. *International Psychogeriatrics, 18*(4), 623–630. https://doi.org/10.1017/S1041610206003887

Husain, G., Thompson, W. F., & Schellenberg, E. G. (2002). Effects of musical tempo and mode on arousal, mood, and spatial abilities. *Music Perception: An Interdisciplinary Journal, 20*(2), 151–171. https://doi.org/10.1525/mp.2002.20.2.151

Janata, P., Tomic, S. T., & Haberman, J. M. (2012). Sensorimotor coupling in music and the psychology of the groove. *Journal of Experimental Psychology. General, 141*(1), 54–75. https://doi.org/10.1037/a0024208

Jones, M. R. (1987). Dynamic pattern structure in music: Recent theory and research. *Perception & psychophysics, 41*(6), 621–634.

Kamil, R. J., & Lin, F. R. (2015). The effects of hearing impairment in older adults on communication partners: a systematic review. *Journal of the American Academy of Audiology, 26*(2), 155–182.

Khalfa, S., Roy, M., Rainville, P., Dalla Bella, S., & Peretz, I. (2008). Role of tempo entrainment in psychophysiological differentiation of happy and sad music? *International Journal of Psychophysiology, 68*(1), 17–26. https://doi.org/10.1016/j.ijpsycho.2007.12.001

Kitwood, T. (1997). *Evaluating dementia care: The DCM method* (7th ed.). Bradford, UK: Bradford Dementia Care Group, University of Bradford.

Klimova, B., Maresova, P., Valis, M., Hort, J., & Kuca, K. (2015). Alzheimer's disease and language impairments: social intervention and medical treatment. *Clinical Interventions in Aging, 10,* 1401–1408. https://doi.org/10.2147/CIA.S89714

Koehne, S., Behrends, A., Fairhurst, M. T., & Dziobek, I. (2016). Fostering social cognition through an imitation- and synchronization-based dance/movement intervention in adults with autism spectrum disorder: A controlled proof-of-concept study. *Psychotherapy and Psychosomatics, 85*(1), 27–35. https://doi.org/10.1159/000441111

Leman, M., Buhmann, J., & Van Dyck, E. (2017). The empowering effects of being locked into the beat of the music. In C. Wöllner (Ed.), *Body, Sound and Space in Music and Beyond: Multimodal Explorations* (pp. 13–28). New York: Routledge.

Lesaffre, M., Moens, B., & Desmet, F. (2017). Monitoring music and movement interaction in people with dementia. In M. Lesaffre, P.-J. Maes, M. Leman (Eds.), *The Routledge Companion to embodied music interaction* (pp. 294–303). New York: Routledge. Retrieved from http://hdl.handle.net/1854/LU-8507317

Marchand, W. R., Lee, J. N., Suchy, Y., Garr., C., Chelune, G., Johnson, S., & Wood, N. (2013). Functional architecture of the cortico-basal ganglia circuitry during motor task execution: Correlations of strength of functional connectivity with neuropsychological task performance among female subjects. *Human Brain Mapping, 34*(5), 1194–1207. https://doi.org/10.1002/hbm.21505

Mas-Herrero, E., Dagher, A., & Zatorre, R. J. (2018). Modulating musical reward sensitivity up and down with transcranial magnetic stimulation. *Nature Human Behaviour, 2*(1), 27–32. https://doi.org/10.1038/s41562-017-0241-z

Merchant, H., Luciana, M., Hooper, C., Majestic, S., & Tuite, P. (2008). Interval timing and Parkinson's disease: Heterogeneity in temporal performance. *Experimental Brain Research, 184*(2), 233–248. https://doi.org/10.1007/s00221-007-1097-7

Narme, P., Clément, S., Ehrlé, N., Schiaratura, L., Vachez, S., Courtaigne, B., Munsch, F., & Samson, S. (2014). Efficacy of musical interventions in dementia: Evidence from a randomized controlled trial. *Journal of Alzheimer's Disease, 38*(2), 359–369. https://doi.org/10.3233/JAD-130893

Nombela, C., Hughes, L. E., Owen, A. M., & Grahn, J. A. (2013). Into the groove: Can rhythm influence Parkinson's disease? *Neuroscience & Biobehavioral Reviews, 37*(10, Part 2), 2564–2570. https://doi.org/10.1016/.neubiorev.2013.08.003

Nozaradan, S., Peretz, I., Missal, M., & Mouraux, A. (2011). Tagging the neuronal entrainment to beat and meter. *Journal of Neuroscience, 31*(28), 10234–10240.

Repp, B. H. (2005). Sensorimotor synchronization: A review of the tapping literature. *Psychonomic Bulletin & Review, 12*(6), 969–992. https://doi.org/10.3758/BF03206433

Repp, B. H., & Su, Y.-H. (2013). Sensorimotor synchronization: A review of recent research (2006–2012). *Psychonomic Bulletin & Review, 20*(3), 403–452. https://doi.org/10.3758/s13423-012-0371-2

Rodger, M. W., & Craig, C. M. (2016). Beyond the metronome: Auditory events and music may afford more than just interval durations as gait cues in Parkinson's disease. *Frontiers in neuroscience, 10*, 272.

Salimpoor, V. N., Benovoy, M., Larcher, K., Dagher, A., & Zatorre, R. J. (2011). Anatomically distinct dopamine release during anticipation and experience of peak emotion to music. *Nature Neuroscience, 14*(2), 257–262. https://doi.org/10.1038/nn.2726

Salimpoor, V. N., Zald, D. H., Zatorre, R. J., Dagher, A., & McIntosh, A. R. (2015). Predictions and the brain: How musical sounds become rewarding. *Trends in Cognitive Sciences, 19*(2), 86–91. https://doi.org/10.1016/j.tics.2014.12.001

Samson, S., Clément, S., Narme, P., Schiaratura, L., & Ehrlé, N. (2015). Efficacy of musical interventions in dementia: methodological requirements of nonpharmacological trials. *Annals of the New York Academy of Sciences, 1337*(1), 249–255. https://doi.org/10.1111/nyas.12621

Särkämö, T., Tervaniemi, M., Laitinen, S., Numminen, A., Kurki, M., Johnson, J. K., & Rantanen, P. (2014). Cognitive, emotional, and social benefits of regular musical activities in early dementia: Randomized controlled study. *The Gerontologist, 54*(4), 634–650. https://doi.org/10.1093/geront/gnt100

Särkämö, T., Laitinen, S., Numminen, A., Kurki, M., Johnson, J. K., & Rantanen, P. (2016). Clinical and demographic factors associated with the cognitive and emotional efficacy of regular musical activities in dementia. *Journal of Alzheimer's Disease, 49*(3), 767–781. https://doi.org/10.3233/JAD-150453

Sebanz, N., Bekkering, H., & Knoblich, G. (2006). Joint action: Bodies and minds moving together. *Trends in cognitive sciences, 10*(2), 70–76.

Sherratt, K., Thornton, A., & Hatton, C. (2004). Emotional and behavioural responses to music in people with dementia: An observational study. *Aging & Mental Health, 8*(3), 233–241. https://doi.org/10.1080/13607860410001669769

Tomasello, M., & Carpenter, M. (2007). Shared intentionality. *Developmental science, 10*(1), 121–125.

Trost, W., Labbé, C., & Grandjean, D. (2017). Rhythmic entrainment as a musical affect induction mechanism. *Neuropsychologia, 96*, 96–110. https://doi.org/10.1016/j.neuropsychologia.2017.01.004

van der Steen, J. T., van Soest-Poortvliet, M. C., van der Wouden, J. C., Bruinsma, M. S., Scholten, R. J., & Vink, A. C. (2017). Music-based therapeutic interventions for people with dementia. *The Cochrane Database of Systematic Reviews, 5*, CD003477. https://doi.org/10.1002/14651858.CD003477.pub3

Wiltermuth, S. S., & Heath, C. (2009). Synchrony and cooperation. *Psychological Science, 20*(1), 1–5. https://doi.org/10.1111/j.1467-9280.2008.02253.x

Zatorre, R. J. (2015). Musical pleasure and reward: Mechanisms and dysfunction. *Annals of the New York Academy of Sciences, 1337*(1), 202–211. https://doi.org/10.1111/nyas.12677

Zatorre, R. J., Chen, J. L., & Penhune, V. B. (2007). When the brain plays music: Auditory-motor interactions in music perception and production. *Nature Reviews Neuroscience, 8*(7), 547–558. https://doi.org/10.1038/nrn2152

Zentner, M., Grandjean, D., & Scherer, K. R. (2008). Emotions evoked by the sound of music: Characterization, classification, and measurement. *Emotion, 8*(4), 494–521. https://doi.org/10.1037/1528-3542.8.4.494

# 3

# Seven Capacities of Music That Underpin its Therapeutic Value in Dementia Care

*Olivia Brancatisano and William Forde Thompson*

Music has been used for health and well-being throughout history and across cultures (Altemüller, Finger, & Boller, 2015). The ancient Greeks employed music to rouse the healthy and heal the sick (Meymandi, 2009), reflecting Plato's famous declaration that music is medicine to the soul (Gfeller, 2005). In contemporary times, there is still widespread agreement that music can have beneficial effects, both physically and psychologically. Engaging with music is a catalyst for brain plasticity (Wan & Schlaug, 2010). It is an enjoyable and effortless way to stimulate an aging brain and develop neural connections. Thus, music may be considered cognitively enriching, keeping the brain active and responsive to the environment. Such benefits may help to rebuild deteriorated neural networks and motivate cognitive and behavioral function in people with dementia.

There is currently no comprehensive model of how music can be used to ameliorate neurological impairment in cognitive, psychosocial, motor and behavioral function. In the past two decades, however, researchers have made considerable progress in understanding the mechanisms underlying the link between music and health (Fujioka et al., 2018; Rickard & McFerran, 2012; Thaut & Hoemberg, 2014). In particular, distinct attributes of music may be linked to specific psychological and physical outcomes (MacDonald, Kreutz, & Mitchell, 2012, pp. 4–6). Drawing from this key insight, Thompson and Schlaug (2015) identified seven capacities of music that seem to be valuable in achieving therapeutic benefits for neurological conditions, as follows: music is engaging, emotional, physical, personal, social, and persuasive, and it affords synchronization. For example, targeted treatments for neurological conditions such as rhythmic auditory stimulation for Parkinson's disease capitalize on the *physical* capacity of music to elicit motor activity that is *synchronous* with a rhythm (Thompson & Schlaug, 2015). Understanding how distinct attributes of music interact with cognitive, psychosocial, behavioral, and motor functions may lead to the development of a comprehensive model of the therapeutic benefits of music for neurological impairment.

In this chapter, we outline the *Therapeutic Music Capacities Model* (TMCM) and explain how each music capacity identified in this model can have therapeutic benefits for individuals with dementia. These capacities enable music to be used in targeted ways across a lifespan to promote health and well-being and can offset the negative consequences of age-related decline.

## Music as a Therapy

In Western society, the professional practice of music therapy has been steadily developing since the 1950s when the first professional societies emerged. Music therapy is a practice in which a qualified music therapist uses music to respond to a client's psychological needs, with emphasis placed on the importance of the therapeutic relationship between the client and therapist. However, the therapeutic benefits of music spread beyond the distinct practice of music therapy to contexts such as music medicine, community music, music education, and everyday uses of music (for review see MacDonald et al., 2012; MacDonald, 2013). For example, the everyday use of music confers benefits through emotion regulation and can be employed in a wide range of settings for a variety of listeners, such as people with dementia in residential care facilities. Music medicine refers to uses of music intervention in contexts such as surgery, palliative care, geriatrics, and neurology and is sometimes practiced by individuals who do not have formal qualifications as a music therapist.

## Music as a Therapeutic Tool for Dementia

The impairments in memory, language, and behavior associated with dementia often result in a deterioration in communication, feelings of agitation, changes in disposition, and loss of identity. Although there is no cure for dementia, pharmacological treatments are available to alleviate certain symptoms, including prescription medications that slow the progress of Alzheimer's disease. Nevertheless, existing treatments can have unwanted side effects, so there is an urgent need to identify alternative therapies to treat symptoms of dementia that minimize negative side effects. Some alternative therapeutic practices, such as cognitive behavior therapy, rely on the individual's ability to communicate in order to engage in the treatment. Unfortunately, in the later stages of dementia, communication skills may be severely impaired or absent. This challenge motivates the use of music, a nonverbal form of communication, as a therapeutic device.

The case of French composer Maurice Ravel sparked interest in how musical abilities can survive despite significant neurodegeneration. It is thought that Ravel suffered from focal cortical dementia (Warren & Rohrer, 2009), and he was unable to communicate or move freely enough to play music towards the end of his life. Remarkably, however, he was still able to compose music. The sparing of Ravel's musical ability is consistent with qualitative and quantitative evidence that musical abilities can sometimes be preserved in individuals with dementia, in spite of widespread impairment in other cognitive-motor functions (Beatty et al.,1988; Crystal, Grober, & Masur, 1989; Fornazzari et al., 2006; for review see Baird & Samson, 2015, Chapters 4, 5, and 8). Researchers have amply documented such preservation of musical abilities in people with Alzheimer's dementia (AD), including the detection of wrong notes in familiar songs (Cuddy & Duffin, 2005), learning new songs by musicians (Cowles, Beatty, Nixon, Lutz, Paulk, Paulk, & Ross, 2003) and nonmusicians (Baird, Umbach, & Thompson, 2017; Prickett & Moore, 1991; Samson, Dellacherie, & Platel, 2009), and the ability to detect emotional meaning in music (Drapeau, Gosselin, Gagnon, Peretz, & Lorrain, 2009). Even people with severe dementia can show preserved emotional responses to music, such as joy (Baird & Thompson, 2018b; Norberg, Melin, & Asplund, 2003). An important implication of these findings is that we may be able to harness these preserved music abilities in a therapeutic way—one that benefits many of the characteristic impairments in dementia and helps individuals maintain a positive quality of life as the disease progresses.

Popular media coverage has highlighted anecdotal cases of the benefits of using music-based treatments in people with dementia. For example, the documentary *Alive Inside* (2014) portrayed music's potential to trigger memories, motivate conversational speech, and instigate dance and other movements in these individuals. The intriguing cases depicted in this documentary are especially significant in illustrating the way that individual properties of music can be isolated or emphasized for specific therapeutic purposes. These anecdotal findings have also been supported by critical reviews on the emotional, cognitive, and behavioral benefits of music and the suggested neural underpinnings of such benefits (e.g., Särkämö, 2017; Särkämö, Altenmüller, Rodríguez-Fornells, & Peretz, 2016). What is it about music that gives it the capacity to nurture health and well-being and to be a source of enjoyment and communicative refuge in the face of severe, progressive neurodegeneration?

## Seven Capacities of Music

Music combines numerous features in a way that is unlike other stimuli. Indeed, one of the distinguishing characteristics of music is that it simultaneously engages a large number of cognitive and motor systems through a multitude of properties, such as melody, harmony, timbre, rhythmic synchronization, phrasing, and emotional expression (e.g., Sihvonen et al., 2017). Manipulation of individual musical features, in turn, can lead to reliable changes in emotional states and cognitive functions (Ilie & Thompson, 2011; Thompson, Schellenberg, & Husain, 2001) and can interact with motivational and reward systems in the brain (Chanda & Levetin, 2013; Salimpoor, Benovoy, Larcher, Dagher, & Zatorre, 2011). Music can also enhance the perceived meaning and significance of a message (Thompson & Russo, 2004), enhance social bonds (Tarr, Launay, & Dunbar, 2014), and enrich one's sense of self (Baird & Thompson, 2018a). In effect, music can be construed as a combination therapy in which distinct elements or capacities interact with various psychological, motor, behavioral and cognitive functions, thereby promoting immediate and, in some cases, lasting therapeutic benefits (for reviews see Altenmüller & Schlaug, 2015; Särkämö et al., 2016; Särkämö, 2017; Sihvonen et al., 2017).

Drawing from a framework proposed by MacDonald et al. (2012) for understanding the therapeutic benefits of music, Thompson and Schlaug (2015) summarised the role of seven core attributes of music that are most strongly linked to its therapeutic value for individuals with neurological impairment. Specifically, they argued that music has therapeutic benefits for neurological impairment because it is persuasive, engaging, emotional, personal, physical, social, and it affords synchronization of movement. These seven capacities combine to make music cognitively enriching for healthy older adults and therapeutic for individuals with dementia, scaffolding core functions such as memory, language fluency, and sense of self.

*Neurocognitive scaffolding* is a natural process by which life-course factors enhance or deplete neural resources, influencing the developmental course of cognition and brain function (Park & Reuter-Lorenz, 2009). For example, engaging in new learning, exercise, and cognitive effort can enhance neural resources, resulting in additional or renewed neural circuitry. Age-related biological consequences, such as white-matter changes, cortical thinning, and dopamine depletion, can contribute to suboptimal scaffolding. In the diseased brain, neuropathology can ultimately lead to a collapse in processes of scaffolding, with consequential decline in multiple functions. However, music, and its broad network of capacities, can engage areas that are typically involved in neurocognitive scaffolding, such as frontal areas that are relatively preserved. By stimulating these areas, music can support this natural process.

The aim of this chapter is to identify the primary qualities of music that contribute to its therapeutic benefits for individuals with neurological impairment and to outline how each quality can be used to target specific neurological problems as indicated by existing evidence. Whilst the list of therapeutic attributes is not exhaustive, the vast majority of observed benefits from music in the literature can be traced to one or more of these seven qualities of music. The model we propose outlines the capacities and the therapeutic outcomes that arise as a result of their therapeutic potential. It also contains hypotheses for how these capacities interact and mutually support one another. For example, the emotional quality of music may be amplified by the personal and social attributes of music, as when an individual listens to an emotional piece of music and reminisces about hearing that music with friends. Typically, multiple capacities and therapeutic benefits co-occur, which is why music is so uniquely powerful.

Figure 3.1 provides an illustration of the *Therapeutic Music Capacities Model* (TMCM). The model begins with some of the contexts in which we may experience music in a therapeutic way (see also MacDonald et al., 2012). These therapeutic contexts are optimally addressed by one or more of the seven capacities that form the core of the model. Arising from the therapeutic use of these capacities are several potential outcomes, including cognitive, psychosocial, motor, and behavioral benefits. Finally, in the smaller dotted boxes are a number of biological and psychological processes that are listed as potentially accounting for the link between the seven individual capacities of music and their beneficial outcomes. The evidence behind these links is explained in further detail in each capacity below.

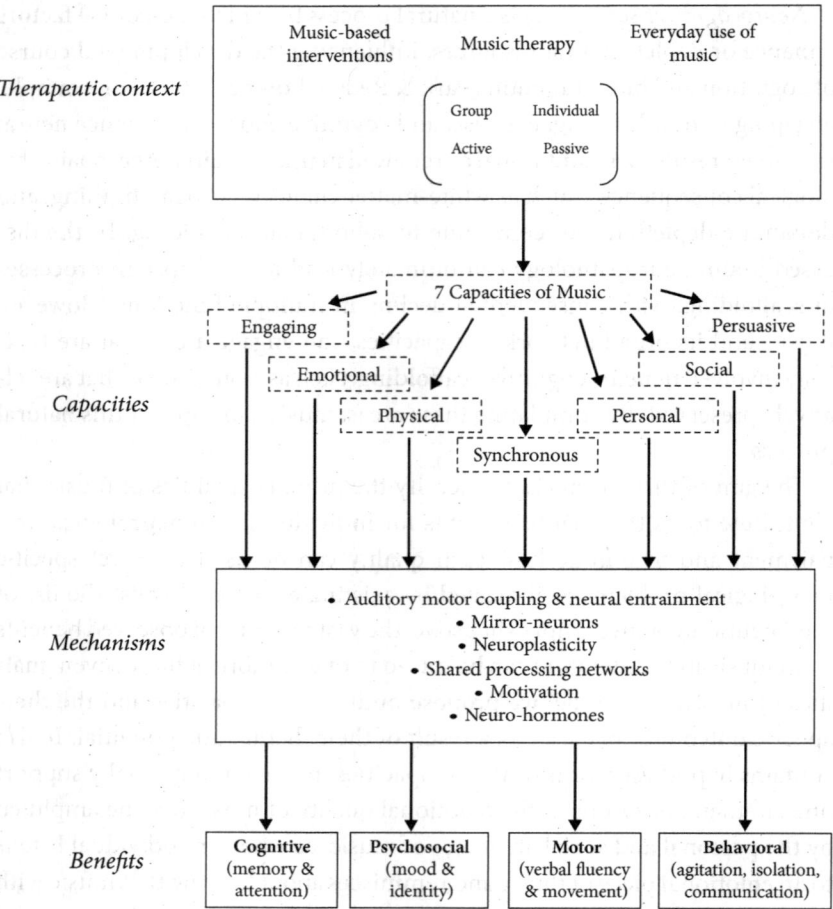

**Figure 3.1.** Therapeutic Music Capacities Model (TMCM).

**Music is *engaging*.** Music engages multiple functions and regions in the brain. Firstly, it engages a range of networks within the brain simultaneously, so it is more of a "whole brain" experience than most other activities. For example, music activates multisensory areas in the frontal, temporal, parietal, and cerebellar regions, as well as emotion and reward centers (Altenmüller & Schlaug, 2015). It also leads to reliable changes in arousal and mood. According to the "mood-arousal" hypothesis, listening to music can enhance certain cognitive functions because of its effects on arousal and mood and associated release of neurochemicals (Thompson et al., 2001). At the initial stages of processing, the temporal lobes process the structural properties of music, such as pitch. As we begin to process the music more deeply, interpreting and experiencing the emotions portrayed, our frontal and mid-brain regions become activated. If we are inclined to move physically to the music, our cerebellar and motor areas are activated. Many brain regions in-volved in music processing, such as the medial frontal and limbic areas, are relatively spared from degeneration in AD (Jacobsen et al., 2015).

Music combines energetic rhythms and catchy melodic phrases to grab at-tention and embed information in a memorable code. Memory for verbal in-formation, such as song lyrics, may be integrated with the detailed pitch and rhythmic information of the music, and this integrated code may be useful for accurate and long-lasting recall (Kilgour et al., 2000; Serafine, Crowder, & Repp, 1984). When background music accompanies the encoding of verbal information, memory performance is improved in older adults (Ferreri et al., 2014).

These engaging features of music make it particularly effective for enhancing memory for verbal information in people with AD. By enriching the encoding of verbal information, less demand is placed on the pre-frontal regions, which are likely to be in decline in AD (Ferreri, Aucouturier, Muthalib, Bigand, & Bugaiska, 2013; Ferreri, Bigand, Bard, & Bugaiska, 2015). Such benefits may arise because music prompts an extensive network of encoding of associated information. By activating widespread networks, representations of music may remain relatively preserved following the dete-rioration of structures dedicated to episodic memory. Research suggests that lyrics are also recognized more easily by people with AD when accompanied by music compared to when they are spoken (Simmons-Stern, Budson, & Ally, 2010; Simmons-Stern et al., 2012). This research has led to the develop-ment of *music mnemonic training* as a way of alleviating memory problems in people with dementia (Gardiner & Thaut, 2014). Music mnemonic training involves pairing certain words with rhythmic or melodic structures and it is listed in the *Neurologic Music Therapy* handbook as a treatment that may be helpful for remembering important everyday information, such as names or lists. Music may facilitate the encoding of verbal material by enhancing neural coherence during new learning (Peterson & Thaut, 2007). Whilst healthy individuals may not need to rely on music-enhanced encoding be-cause they have intact cortical structures for memory, the mnemonic benefits provided by music may be necessary for people with AD, (Kilgour, Jacobsen, & Cuddy, 2000). The engaging capacity of music to activate brain regions that are spared from degeneration, as well as its ability to enrich information which places less demand on degenerating brain regions, provides a unique opportunity to harness music as a tool for rehabilitation in dementia (for review, see Särkämö, 2017).

**Music is *emotional*.** One of most significant purposes of music is to convey emotional meaning. Music can elicit complex emotions ranging from nostalgia and yearning to empowerment and joy. The association between emotion and music is observed across cultures, perhaps because human emotions spontaneously track basic acoustic features in the environment, such as intensity, pitch height, and rate (Balkwill & Thompson, 1999; Balkwill, Thompson, & Matsunaga, 2004; Fritz et al., 2009; Ilie & Thompson, 2006; Ma & Thompson, 2015). Such findings suggest that the tendency to perceive emotion in music may be hardwired. Certain musical events, such as an increase in intensity, can lead to changes in several physiological measures indicative of emotional arousal (Krumhansl, 1997; Olsen & Stevens, 2013), indicating that music can have a direct impact on the automatic nervous system, endocrine system, and para-limbic brain structures (Blood & Zatorre, 2001; Koelsch, 2010, 2014; Lundqvist, Carlsson, Hilmersson, & Juslin, 2008).

Music's ability to alter our emotions makes it an effective tool for mood regulation (Moore, 2013). Among elderly individuals, music listening is a common leisure activity used for pleasure, mood regulation, and relaxation— a practice that has positive effects on well-being (Laukka, 2007). A large proportion of the research on music, emotion and dementia has been conducted in individuals with AD. These individuals show intact processing of musical emotions relative to those with other types of dementia, such as semantic dementia and behavioral variant frontotemporal dementia, a group known as the frontotemporal lobar degenerations (Hsieh, Hornberger, Piguet, & Hodges, 2012; Omar et al., 2011, see Chapter 5). The recognition of musical emotions is spared in people with AD, even though they may be impaired in recognition of emotions from facial cues (Drapeau et al., 2009). Further, people with AD demonstrate a comparable ability with healthy individuals to use musical elements, such as tempo and mode, to interpret musical emotions (Gagnon et al., 2009). Such findings support the use of music in dementia care, particularly in AD. The way in which music can heighten emotions can be utilized to reduce apathy in people with moderate to severe AD. Apathy is one of the most common neuropsychiatric symptoms in AD and mild cognitive impairment, manifesting as a loss of interest and engagement, and a lack of or blunted emotional responses (Mortby, Maercker,

& Forstmeier, 2012). Receptive music-based treatments have been shown to significantly decrease apathy (Massaia et al., 2018; Tang et al., 2018). In particular, one study found that listening to live music was effective at reducing apathy and improving engagement, over and above prerecorded music (Holmes, Knights, Dean, Hodkinson, & Hopkins, 2006). Furthermore, in music-therapy interventions, participants showed a decrease in apathy and an increase in smiling behaviors compared to a control intervention of standard care (Raglio et al., 2008, see Chapter 10).

Importantly, in individuals with AD, music also plays an important role in re-gaining access to emotions through autobiographical memories. The capacity of music to act as a scaffold for autobiographical memories is spared in people with AD. Music can elicit memories from a particular time of one's life and remembering significant events can often elicit the same emotions that were experienced at the time, for example the joy of seeing a loved one after a long separation (Juslin, Liljeström, Västfjäll, & Lundqvist, 2010; Juslin & Västfjäll, 2008). A number of researchers have investigated the autobiographical memories that are evoked while listening to music, also called *music-evoked autobiographical memories* (MEAMs). Research on the incidence of MEAMs in older adults has established that people with AD may experience just as many MEAMs as healthy elderly individuals (Baird, Brancatisano, Gelding, & Thompson, 2018; Cuddy, Sikka, & Vandstone, 2015). This finding illustrates that, despite a dramatic decline in most forms of memory, autobiographical recollections evoked by music remain relatively preserved. Furthermore, the MEAMs that are triggered by music are richly emotional (Cuddy, Sikka, Silveira, Bai, & Vanstone, 2017; El Haj, Fasotti, & Allain, 2012). Interestingly, the memories evoked by music tend to contain more emotional content, and are more positively valenced, than those evoked in silence (Cuddy et al., 2017; El Haj et al., 2012). Cuddy et al. (2017) found that compared to younger adults, the MEAMs produced by older adults with AD included more positive content. The participants with AD also self-rated these memories as more positive and less negative. This finding is known as the *positivity effect* and describes how later in life adults become more positive about life. This implies that the effects of music can benefit people with AD by eliciting positive memories and, subsequently, inducing an optimistic state of mind. In individuals with AD, feelings can persist even with no explicit recall of a memory for that emotional event (Guzmán-Vélez, Feinstein, & Tranel, 2014).

Music is *physical*. It is often hard to separate the experiences of music and movement. In many cultures, one without the other is inconceivable. This strong association illustrates that music is inherently a physical stimulus. Firstly, musical sounds are themselves the outcome of physical movements of the human body (e.g., singing, clapping, blowing, plucking). Secondly, when we hear certain types of music we get a strong urge to move our body to this music. Motor areas within the brain, normally associated with explicit bodily movement, are also activated from simply listening to rhythmical music even when no movement is generated (Chen, Penhune, & Zatorre, 2008). In this way, music is strongly associated with physical movement (e.g., dancing, marching, figure skating, and ballet).

The positive effects of physical exercise on aging populations are increasingly recognized and have profound implications for the well-being and quality of life of people with, or at risk of, dementia. Engaging in physical exercise has been known to slow the onset of dementia and cognitive impairment (Laurin, Verreault, Lindsay, MacPherson, Rockwood, 2001; Larson et al., 2006; for review see Carvalho, Rea, Parimon, & Cusack, 2014). Furthermore, in a longitudinal review over 5 years, engagement in leisure activities such as musical instrument playing and dancing reduced the risk of dementia (Verghese et al., 2003). Pairing music and movement may improve physical abilities important to maintain in the elderly, such as motor and tactile functions, and posture (Kattenstroth, Kalisch, Holt, Tegenthoff, & Dinse, 2013). Specifically, aerobic exercise has been shown to increase hippocampal volume in healthy older adults (Erickson et al., 2011) and older women with mild cognitive impairment (Ten Brinke et al., 2015). Erickson et al. (2011) also noted that the increase in hippocampal volume in healthy elderly may result in improvements in spatial memory. Furthermore, exercise and its associated benefits for memory may be accompanied by an increase in the production of brain-derived neurotrophic factor, which mediates neurogenesis (Erickson et al., 2011). Thus, pairing music and aerobic exercise may maximize therapeutic benefits, extending beyond direct physical outcomes to enhancing cognition.

Preserved motor responses to music have been documented anecdotally over the past decade with documentaries and YouTube videos showing footage of elderly individuals, who, seemingly bound to their chair or walking frames, suddenly get up and move to music with a strong beat. Similarly, some individuals who are nonresponsive will still tap a foot or finger when they hear a rhythm that reignites their sense of movement.

From these anecdotal cases it seems that during the progression of dementia, motor responses to music may endure longer than responses to other stimuli, such that even when individuals suffer cognitive decline, they may still able to move to music.

Therapeutic interventions have capitalized on the physically motivating properties of music by pairing music and movement in programs, subsequently benefitting cognitive function, mood, and behavior. A recent randomized control trial compared the effectiveness of a "Music with Movement" group intervention with a control group which received standard care (Cheung, Lai, Wong, & Leung, 2016). The intervention involved participants moving their body to familiar music for 30 minutes, twice a week over 4 weeks. Across this time frame, there was a greater decrease in agitated behaviors for the music and movement group than for the control group. It is difficult to tease apart the relevant contributions of physical movement, familiarity with the music, and the social setting of the task. However, research suggests that these factors may interact with and support one another.

Another study assessed cognitive outcomes in healthy elderly after completing an intervention of exercise with or without music for one hour per week over the course of a year (Satoh et al., 2014). Only the individuals who exercised with music exhibited improvements in visuospatial function and cognitive function (as measured by the mini-mental state examination), demonstrating that the combination of music and exercise is more beneficial than exercise alone. One reason for the effectiveness of such an intervention could be the association between music and physical movement, which increases our motivation to move with the music. When we are highly motivated to participate in a therapeutic intervention, it is more likely to confer benefits. Moreover, physical responses to music also tend to be more enjoyable than exercise for its own sake, such that the benefits of exercise are combined with additional benefits of enhanced mood.

**Music affords *synchronization*.** Music not only motivates physical action, but it affords the synchronous movements of these actions in time with the music itself and with other individuals. The synchronization of music with movement and other individuals may have served an evolutionary purpose to enhance group cohesion and coordination (Huron, 2001, Phillips-Silver, 2009). For example, bonding occurs when infants are rocked in time to lullabies sung by their mothers (Trehub, Unyk, & Trainor 1993; Trehub, 2003), and choirs and orchestras foster group cohesion through singing and playing music in time with one another (Huron, 2001). A number of therapeutic benefits have been observed when groups of individuals synchronize their movements with one another (e.g., dancing together, or clapping together). For example, in synchronous dancing or drumming, there is a release of endorphins and neurochemicals such as oxytocin, which are responsible for feelings of social bonding and empathy (Domes et al., 2007; Tarr et al., 2014) and trust (Kosfeld, Heinrichs, Zak, Fischbacher, & Fehr, 2005). A reduction in pain has also been noted after synchronizing movements to music in a group scenario rather than merely listening to music (Dunbar et al., 2012).

People with dementia remain able to synchronize with music through entrainment (see Chapter 2). Synchronizing with music affords therapeutic motor, cognitive, and behavioral benefits in people with dementia. Firstly, the tendency to move in response to music may assist in the maintenance of motor functions in AD and in learning new movement sequences (Moussard, Bigand, Belleville, & Peretz, 2014). Further, when synchronous drum playing is encouraged in music therapy among patients with dementia, it may lead to increased awareness and verbal and nonverbal communication as well as a decrease in agitation (Allen-Williams, 2013). Synchronizing to an external rhythm is used as a technique in music therapy to reduce negative behavioral symptoms common to people with dementia, such as agitation. A music therapist may use a simple rhythmic pattern on a drum that matches or emulates the patient's agitated rhythmic pattern, helping to synchronize an external pattern to their internal pattern (see Chapter 11). The rhythmic pattern is seen as a way to empathetically connect with the person with dementia and subsequently give a sense of regulation and security, reducing the disordered behaviors (Allen-Williams, 2013). A plausible mechanism underlying our ability to synchronize with music is *neural entrainment*—the repetitive and synchronous firing of neurons in response to a clear rhythmic pulse, such that the brain's electrical activity aligns with the beat of the music (Thaut, Kenyon, Schauer, & McIntosh, 1999; Dowelling & Poppel, 2015, see Chapter 2).

The combination of physical and synchronous attributes of music creates a uniquely powerful therapeutic benefit. In a recent study, Shimizu, Umemura, Matsunaga, and Hirai (2017) tested whether *music movement therapy* improved mobility and executive function in people with mild cognitive impairment. The music movement therapy involved pairing music with movement and playing a familiar Japanese rhythmical instrument. The music movement therapy was compared to a control therapy that completed the same movements as those in the music with movement group, but in the absence of music. Whilst both groups improved in several of the physical measures, such as flexibility, mobility, gait, and muscle endurance, the music movement therapy resulted in a greater improvement in test scores on the frontal assessment battery (a tool for the assessment of frontal lobe functions such as abstraction, fluency, and impulsivity) and a larger increase in the level of cerebral blood flow to the prefrontal cortex as measured by near-infrared spectroscopy. These findings suggest that synchronization, through playing the rhythmical musical instruments in time with one other, stimulated neural activation in the prefrontal cortex with corresponding improvement in cognitive functioning.

**Music is *personal*.** One of the most powerful features of the musical experience is the personal connection that we are able to build with it. During adolescence and early adulthood, music plays an important role in establishing our identity (North & Hargreaves, 1999). We often listen to music throughout our lives to reinforce our sense of self. When this same music is heard later in life, it often evokes a sense of familiarity. Songs with autobiographical significance can both reaffirm the sense of self or "awaken" those with dementia (Baird & Thompson, 2018a, b). In some instances, spouses have sung personally significant songs to their partner with severe dementia as a way of reconnecting with them. This process has often prompted the person with dementia to "come back" and finally recognize their partner after periods of misidentifying them as a stranger (Baird & Thompson, 2018a).

The sense of familiarity that comes with personal music can help people with dementia become more oriented with a new environment or maximize their sense of familiarity in a current one (Son, Therrien, & Whall, 2002). Playing or singing personally preferred music to people with dementia is a technique often employed to reduce agitation and anxiety and induce a sense of calm (e.g., Ridder, Stige, Qvale, & Gold, 2013 and see Chapter 11). *Music-Therapeutic Caregiving* is a method of communication in which a caregiver sings favorite songs to people with advanced dementia, especially during daily activities such as dressing, meal time, and medication time. Evidence suggests that Music-Therapeutic Caregiving can lead to improvements in patient compliance during standard care activities (Brown, Gotell, & Ekman, 2001). Gerdner (2000) demonstrated the effect of using personally preferred music, as opposed to relaxing classical music, to decrease levels of agitation, highlighting the importance of the personal nature of music. Furthermore,

Särkämö et al. (2014) compared two 10-week treatments of song-listening and singing in people in the early stages of dementia. Individuals in the song-listening group were asked to identify songs that were most autobiographically relevant to them. Results showed that listening to the personal songs resulted in a greater level of calm and a reduction in agitation. Intact musical memory in AD may explain why familiar music can be a successful therapeutic tool in this population (Cuddy et al., 2015 and see Chapter 4).

Music that is heard repeatedly during significant or pivotal times in our personal development, through persistent association, eventually seems to *signify* that time of life. In this way, music is a powerful stimulus for reminiscence. The "reminiscence bump" is a particularly rich period of memories and songs that evoke significantly more memories than those from other decades (Baird et al., 2018). In a study by Dassa and Amir (2014), participants with middle- to late-stage AD took part in eight sessions of music therapy in which they sang popular songs in a group setting and reminisced. Songs that were personally significant to the participants, such as those that related to their social and national identity, had the greatest effect on memory retrieval, as assessed by content analysis. Additionally, spontaneous speech increased in the conversations that were held after the group singing. Further analyses suggested that certain musical features interacted with the ability to participate in memory-rich conversations. For example, the marching tempo of songs relating to national identity seemed to induce patriotic memories and feelings of social unity. Conversely, songs with a triple meter and repetitive melodic and rhythmic patterns often seemed to evoke memories associated with social contexts, such as "sitting around a campfire together." In short, songs that are both personal and contain evocative structural properties have a powerful capacity to reinforce an individual's sense of self, engaging patients and evoking memory-rich, elaborative speech. In this way, the personal nature of music may be used to stimulate verbal communication in people with dementia.

**Music is *social*.** One of the most significant challenges associated with dementia is social isolation, which can occur following the decline in behavioral and verbal functions. Music can play an important role in addressing this problem. Music acts as a catalyst for bringing people together and enhancing a group experience (for review see Huron, 2001). It is shared in times of celebration, ritual, and mourning to produce a sense of cohesion among individuals.

Unlike turn-taking social activities such as talking, making music can be done with everyone simultaneously expressing themselves whilst each individual is still heard. The social dimension of music may enhance its other benefits. Experiencing music as a group can enhance *engagement* by encouraging eye contact and other forms of social interactions with one another. It can heighten *emotional* experiences associated with social interaction by increasing "feel-good" hormones such as oxytocin (Keeler et al., 2015). It can deepen the *personal* nature of music by encouraging discussion of individual experiences with one another and creating a group identity. Experiencing music in a social setting also allows for the elements of the *physical* nature of music and *synchronization* to be enhanced via imitation, thus building a strong sense of cohesion. Overy and Molnar-Szakacs (2009) suggest this is why music is such a dominating stimulus in human societies and that imitation, synchronization, and shared experience are "key elements" for successful therapeutic programs.

Throughout life, especially during the later stages, music acts as a powerful means of social interaction (Juslin & Sloboda, 2011). For people with dementia, this is particularly helpful for reducing isolation. There is strong evidence that group singing can benefit those with dementia (see Chapter 13). Improvements have been demonstrated in cognitive function, behavior, mood, and well-being after participating in group singing and music activities (Name et al., 2014; Sakamoto et al., 2013; Särkämö et al., 2014). One study by Osman, Tickler, and Schneider (2016) examined the effect that a particular group singing program, *Singing for the Brain*, had on people with AD and their carers. Semi-structured interviews were gathered and thematically analyzed to determine the impact of the intervention on both those with dementia and carers. A number of positive outcomes were observed in people with dementia after group singing, such as feelings of inclusivity, enjoyment of the sessions, and self-reported improvement in mood and memory. Importantly, many participants also indicated that group singing helped them to accept and cope with their disease.

Evidence also suggests that group music therapy has other psychological and behavioral benefits for people with dementia, such as reducing anxiety and agitation (e.g., Ing-Randolph, Phillips, & Williams, 2015; Ledger & Baker, 2007; Lin et al., 2011; Raglio et al., 2008; Särkämö et al., 2014), though some studies have yielded mixed results (Solé, Mercadal-Brotons, Galati, & De Castro, 2014; Raglio et al., 2015). For example, Raglio et al. (2015) reported music therapy demonstrated no difference to standard care in the reduction of neuropsychiatric symptoms such as delusions, anxiety, and disinhibition. Such discrepancies in results may reflect methdological differences, for example in group size and the severity of dementia (see Part 3).

Särkämö et al., (2016) also found differences in the effects of group singing compared to group music listening interventions for people with mild to moderate dementia. There was a greater reduction in behavioral symptoms of depression, such as agitation and loss of interest in the music listening group, whereas in the singing group, there was a greater improvement in physical symptoms of depression, such as increased energy and weight control. These findings are consistent with results obtained for healthy elderly adults, who reported higher ratings of physical quality of life after partaking in a group choir (Johnson, Louhivuori, & Siljander, 2017). Särkämö and colleagues suggest that the differing effects that group music listening and singing may have can be attributed to their unique nature: "listening to music and the associated reminiscence can be more calming and relaxing, whereas engaging in joint singing can be more energizing, refreshing, and stress-reducing" (p. 440).

**Music is *persuasive*.** Lastly, music has the capacity to influence us. It has been used historically as a tool to reinforce, change, or inspire beliefs. Whether it be to assist in healing rituals, religious ceremonies, election campaigns, or a political movement such as the peace movement in the 1960's, music is a highly effective device for the manipulation of belief systems. Music is particularly successful at belief manipulation because its effects are often implicit. Music cannot instruct listeners in what they should believe; rather, it interacts with belief systems by evoking powerful emotions that highlight the significance and perceived gravitas of any accompanying media, such as lyrical content. For example, lyrics are perceived to be more persuasive if they are paired with music than if they are decoupled from the music because the emotional impact of music is conflated with any accompanying messages— a process referred to as "trickle-down meaning" by Thompson and Russo (2004). Therapeutic song writing is a music-therapy practice which draws upon this idea, whereby lyrics are created by the patient with dementia or caregiver and paired with music to address psychosocial and emotional needs (Baker, Wigram, Stott, & McFerran, 2008). By embedding content related to dementia in a musical context, perspectives or beliefs are able to be reappraised and reframed (Baker, 2015). Therapeutic song writing has also been shown to motivate participants with dementia to engage in regular treatment and instill empowerment and agency (Baker & Stretton-Smith, 2018). Further, this practice has been described as highly enjoyable by people with dementia. In any therapeutic setting, enjoyment tends to increase motivation for participating in treatment and optimism for its impact on quality of life (Baker & Ballantyne, 2013). Thus, music is persuasive not only in the sense that it highlights and supports accompanying ideas, but also because the sheer enjoyment it stimulates leads to an optimistic outlook.

## Conclusion

The Therapeutic Music Capacities Model describes seven enriching capacities of music that play a significant role in its use as a therapeutic treatment for symptoms of dementia. Each of these seven individual musical attributes has potential to make a unique contribution in the care of people with dementia. However, it is rarely the case that these attributes occur independently in music activities. Indeed, one of the reasons music is so powerful as a treatment tool for dementia is that these seven attributes are combined in music and have cumulative benefits. In this way, music engagement encompasses new learning, exercise, and cognitive engagement that optimally supports neurocognitive scaffolding.

In a research setting, it is challenging to examine each attribute separately, as they are difficult to disentangle from one another. Attempts to isolate each attribute may result in simplified stimuli that are not representative of the richness of real-world music experiences. Despite this scientific challenge, there is reason to believe that each of the seven qualities is associated with particular mechanisms of action. These mechanisms may interact with each other in ways that are not yet fully understood. Understanding the mechanisms underlying the therapeutic value of these capacities and their interplay will require concerted research efforts. The Therapeutic Music Capacities Model provides an initial step towards dissecting current treatments to determine why *music* is such a powerful tool in this context. The ultimate goal of the framework is to improve the quality of music-based treatments for dementia.

## References

Allen-Williams, A. (2013). Does rhythm in music therapy have an organising effect on the agitated behaviours of people with a diagnosis of dementia? An investigation into music therapists' current practice. *British Journal of Music Therapy, 27*(1), 32–51. doi:10.1177/135945751302700104

Altenmüller, E., Finger, S., & Boller, F. (2015). Music, neurology and neuroscience: Historical connections and perspectives. *Progress in Brain Research, 216,* 2–422. Retrieved from https://www.sciencedirect.com/bookseries/progress-in-brain-research/vol/216.

Altenmüller, E., & Schlaug, G. (2015). Apollo's gift: New aspects of neurologic music therapy. *Progress in Brain Research, 217,* 237–252. doi:10.1016/bs.pbr.2014.11.029

Baird, A., Brancatisano, O., Gelding, R., & Thompson, W. F. (2018). Characterization of music and photograph evoked autobiographical memories in people with Alzheimer's disease. *Journal of Alzheimer's Disease, 66*(2) doi:10.3233/JAD-180627

Baird, A., & Samson, S. (2015). Music and dementia. In E. Altenmüller, F. Boller, & S. Finger (Eds.), *Music, Neurology, and Neuroscience: History and Modern Perspectives (Progress in Brain Research Series).* London, UK: Elsevier.

Baird, A., & Thompson, W. F. (2018a). The impact of music on the self in dementia. *Journal of Alzheimer's Disease, 61*(3),827–841. doi:10.3233/JAD-170737

Baird, A., & Thompson, W. F. (2018b). When music compensates language: A case study of severe aphasia in dementia and the use of music by a spousal caregiver. *Aphasiology.* doi:10.1080/02687038.2018.1471657

Baird, A., Umbach, H., & Thompson, W. F. (2017). A non-musician with severe Alzheimer's dementia learns a new song. *Neurocase, 23*(1), 36–40. doi:10.1080/13554794.2017.1287278

Baker, F. A. (2015). *Therapeutic songwriting: Developments in theory, methods, and practice.* London: Palgrave MacMillan

Baker, F., & Ballantyne, J. (2013). "You've got to accentuate the positive": Group songwriting to promote a life of enjoyment, engagement and meaning in aging Australians. *Nordic Journal of Music Therapy, 22*(1), 7–24. doi.org/10.1080/08098131.2012.678372

Baker, F., & Stretton-Smith, P. A. (2018). Group therapeutic songwriting and dementia: Exploring the perspectives of participants through interpretative phenomenological analysis. *Music Therapy Perspectives, 36*(1), 50–66. doi:10.1093/mtp/mix016

Baker, F., Wigram, T., Stott, D., & McFerran, K. (2008). Therapeutic songwriting in music therapy: Part I. Who are the therapists, who are the clients, and why is songwriting used? *Nordic Journal of Music Therapy, 17*(2), 105–123.

Balkwill, L. L., & Thompson, W. F. (1999). A cross-cultural investigation of the perception of emotion in music: Psychophysical and cultural cues. *Music Perception: An Interdisciplinary Journal, 17*(1), 43–64, doi:10.2307/40285811

Balkwill, L. L., Thompson, W. F., & Matsunaga, R. (2004). Recognition of emotion in Japanese, Western, and Hindustani music by Japanese listeners. *Japanese Psychological Research, 46,* 337–349. doi:10.1111/j.1468-5584.2004.00265.x

Beatty, W. W., Zavadil, K. D., Bailly, R. C., & Rixen, G. J., Zavadil, L. E., Farnham, N., & Fisher, L. (1988). Preserved musical skill in a severely demented patient. *International Journal of Clinical Neuropsychology, 10*(4), 158–164.

Blood, A. J., & Zatorre, R. J. (2001). Intensely pleasurable responses to music correlate with activity in brain regions implicated in reward and emotion. *Proceedings of the National Academy of Sciences of the United States of America, 98*(20), 11818–11823. doi:10.1073/pnas.191355898

Brown, S., Gotell, E., & Ekman, S. L. (2001). Singing as a therapeutic intervention in dementia care. *Journal of Dementia Care, 9*(4), 33–37.

Carvalho, A., Rea, I. M., Parimon, T., & Cusack, B. J. (2014). Physical activity and cognitive function in individuals over 60 years of age: A systematic review. *Clinical Interventions in Aging, 9,* 661–682. doi:10.2147/CIA.S55520

Chanda, M. L., & Levitin, D. J. (2013). The neurochemistry of music. *Trends in Cognitive Sciences, 17*(4), 179–193. doi:10.1016/j.tics.2013.02.007

Chen, J. L., Penhune, V. B., & Zatorre, R. J. (2008). Listening to musical rhythms recruits motor regions of the brain. *Cerebral Cortex, 18*(12), 2844–2854. doi:10.1093/cercor/bhn042

Cheung, D. S. K., Lai, C. K. Y., Wong, F. K. Y., & Leung, M. C. P. (2016). The effects of the music-with-movement intervention on the cognitive functions of people with moderate dementia: A randomized controlled trial. *Aging & Mental Health, 22*(3), 306–315. doi.org/10.1080/13607863.2016.1251571

Cowles, A., Beatty, W. W., Nixon, S. J., Lutz, L. J., Paulk, J., Paulk, K., & Ross, E. D. (2003). Musical skill in dementia: A violinist presumed to have Alzheimer's disease learns to play a new song. *Neurocase, 9,* 493–503. doi:10.1076/neur.9.6.493.29378

Crystal, H. A., Grober, E., & Masur, D. (1989). Preservation of musical memory in Alzheimer's disease. *Journal of Neurology, Neurosurgery, and Psychiatry, 52*(12), 1415–1416.

Cuddy, L. L., & Duffin, J. (2005). Music, memory, and Alzheimer's disease: Is music recognition spared in dementia, and how can it be assessed? *Medical Hypotheses, 64*(2), 229–235.

Cuddy, L. L., Sikka, R., Silveira, K., Bai, S., & Vanstone, A. (2017). Music evoked autobiographical memories in Alzheimer's disease: Evidence for a positivity effect. *Cogent Psychology, 4,* 1277578. doi:10.1080/23311908 2016.1277573

Cuddy, L. L., Sikka, R., & Vandstone, A. (2015). Preservation of musical memory and engagement in healthy aging and Alzheimer's disease. *Annals of the New York Academy of Sciences, 1337,* 223–231. doi:10.1111/nyas.12617

Dassa, A., & Amir, D. (2014). The role of singing familiar songs in encouraging conversation among people with middle to late stage Alzheimer's disease. *Journal of Music Therapy, 51*(2), 131–153. doi:10.1093/jmt/thu007

Doelling, K. B., & Poeppel, D. (2015). Cortical entrainment to music and its modulation by expertise. *Proceedings of the National Academy of Sciences of the United States of America, 112*(45), E6233–6242. doi:10.1073/pnas.1508431112

Domes, G., Heinrichs, M., Michel, A., Berger, C., & Herpertz, S.C. (2007). oxytocin improves "mind-reading" in humans. *Biological Psychiatry, 61*(6), 731–733.

Drapeau, J., Gosselin, N., Gagnon, L., Peretz, I., & Lorrain, D (2009). Emotional recognition from face, voice, and music in dementia of the Alzheimer type. *Annals of the New York Academy of Sciences, 1169,* 342–345. doi:10.1111/j.1749-6632.2009.04768.x

Dunbar, R.I.M., Kaskatis, K., MacDonald, I., & Barra, V. (2012). Performance of music elevates pain threshold and positive affect: Implications for the evolutionary function of music. *Evolutionary Psychology, 10*(4), 688–702. doi:10.1177/147470491201000403

El Haj, M., Fasotti, L., & Allain, P. (2012). The involuntary nature of music-evoked autobiographical memories in Alzheimer's disease. *Consciousness and Cognition, 21,* 238–246. doi:10.1016/j.concog.2011.12.005

El Haj, M., Postal, V., & Allain, P. (2012). Music enhances autobiographical memory in mild Alzheimer's disease. *Educational Gerontology, 38,* 30–41. doi:10.1080/03601277.2010.515897

Erickson, K. I., Voss, M. W., Prakash, R. S., Basak, C., Szabo, A. Chaddock, L., . . . Kramer, A. F. (2011). Exercise training increases size of hippocampus and improves memory. *Proceedings of the National Academy of Sciences of the United States of America, 108*(7), 3017–3022. doi:10.1073/pnas.1015950108

Ferreri, L., Aucouturier, J.-J., Muthalib, M., Bigand, E., & Bugaiska, A. (2013). Music improves verbal memory encoding while decreasing prefrontal cortex activity: An fNIRS study. *Frontiers in Human Neuroscience, 7,* 779. doi:10.3389/fnhum.2013.00779

Ferreri, L., Bigand, E., Bard, P., & Bugaiska, A. (2015). The influence of music on prefrontal cortex during episodic encoding and retrieval of verbal information: A multichannel fNIRS study. *Behavioural Neurology, 2015,* Article ID 707625, 11 pages, 2015. doi:10.1155/2015/707625

Ferreri, L., Bigand, E., Perrey, S., Muthalib, M., Bard, P., & Bugaiska, A. (2014). Less effort, better results: How does music act on prefrontal cortex in older adults during verbal encoding? An fNIRS Study. *Frontiers in Human Neuroscience, 8*, 301. doi:10.3389/fnhum.2014.00301

Fornazzari, L., et al. (2006). Preservation of episodic musical memory in a pianist with Alzheimer disease. *Neurology, 66* (4), 610–611. doi:10.1212/01.WNL.0000198242. 13411.FB

Fritz, T., Jentschke, S., Gosselin, N., Sammler, D., Peretz, I., Turner, R., . . . Koelsch, S. (2009). Universal Recognition of Three Basic Emotions in Music. *Current Biology, 19*(7), 573–576. doi:10.1016/j.cub.2009.02.058

Fujioka, T., Dawson, D. R., Wright, R., Honjo, K., Chen, J. L., Chen, J. J., . . . Ross, B. (2018), The effects of music-supported therapy on motor, cognitive, and psychosocial functions in chronic stroke. *Annals of the New York Academy of Sciences, 1423*, 264–274. doi:10.1111/nyas.13706

Gagnon, L., Peretz, I., & Fülöp, T. (2009). Musical structural determinants of emotional judgments in dementia of the Alzheimer type. *Neuropsychology, 23*(1), 90–97. doi:10.1037/a0013790

Gardiner, J. C., & Thaut, M. (2014). Musical Mnemonics Training (MMT). In: M.H. Thaut & H.Volker (Eds.), *Handbook of Neurologic Music Therapy* (pp. 294–310). Oxford, UK: Oxford University Press.

Gerdner, L. A. (2000). Effects of individualized versus classical "relaxation" music on the frequency of agitation in elderly persons with Alzheimer's disease and related disorders. *International Psychogeriatrics, 12*(1), 49–65.

Gfleller, K. (2005). Music as a therapeutic agent: Sociocultural perspectives. In R. Unkefer & M. Thaut (Eds.), *Music therapy in the treatment of adults with mental disorders: Theoretical bases and clinical interventions*, pp. 60–67. Gilsum NH: Barcelona.

Guzmán-Vélez, E., Feinstein, J. S., & Tranel, D. (2014). Feelings without memory in Alzheimer disease. *Cognitive and behavioral neurology: Official journal of the Society for Behavioral and Cognitive Neurology, 27*(3), 117–29.

Holmes, C., Knights, A., Dean, C., Hodkinson, S., & Hopkins, V. (2006). Keep music live: Music and the alleviation of apathy in dementia subjects. *International Psychogeriatrics, 218*(4):623–630.

Hsieh, S., Hornberger, M., Piguet, O., & Hodges, J. (2012). Brain correlates of musical and facial emotion recognition: Evidence from the dementias. *Neuropsychologia, 50*(8), 1814–1822. doi:10.1016/j.neuropsychologia.2012.04.006

Huron, D. (2001). Is music an evolutionary adaptation? *Annals of the New York Academy of Sciences, 930*, 43–61.

Ilie, G., & Thompson, W. F. (2006). A comparison of acoustic cues in music and speech for three dimensions of affect. *Music Perception: An Interdisciplinary Journal, 23*(4), 319–330. doi:10.1525/mp.2006.23.4.319

Ilie, G., & Thompson, W. F. (2011). Experiential and cognitive changes following seven minutes exposure to music and speech. *Music Perception: An Interdisciplinary Journal, 28*(3), 247–264. doi:10.1525/mp.2011.28.3.247

Ing-Randolph, A.R., Phillips, L.R., Williams, A.B. (2015). Group music interventions for dementia-associated anxiety: A systematic review. *International Journal of Nursing Studies, 52*(11), 1775–1784. doi:10.1016/j.ijnurstu.2015.06.014.

Jacobsen, J.-H., Stelzer, J., Fritz, T. H., Chételat, G., La Joie, R., & Turner, R. (2015). Why musical memory can be preserved in advanced Alzheimer's disease. *Brain, 138*(8), 2438–2450. doi:10.1093/brain/awv135

Johnson, J. K., Louhivuori, J., & Siljander, E. (2017). Comparison of well-being of older adult choir singers and the general population in Finland: A case-control study. *Musicae Scientiae, 21*(2):178–194. doi:10.1177/1029864916644486

Juslin, P. N., Liljeström, S., Västfjäll, D., & Lundcvist, L. O. (2010). How does music evoke emotions? Exploring the underlying mechanisms. In P. N. Juslin & J. A. Sloboda (Eds.), *Series in affective science. Handbook of music and emotion: Theory, research, applications* (pp. 605–642). New York,: Oxford University Press.

Juslin, P. N., & Sloboda, J. (Eds.). (2011). *Handbook of music and emotion: Theory, research, applications.* New York: Oxford University Press.

Juslin, P. N., & Västfjäll, D. (2008). Emotional responses to music: The need to consider underlying mechanisms. *Behavioral and Brain Sciences, 31*(5), 559–575; doi:10.1017/S0140525X08005293

Kattenstroth, J. C., Kalisch, T., Holt, S., Tegenthoff, M., & Dinse, H. R. (2013). Six months of dance intervention enhances postural, sensorimotor, and cognitive performance in elderly without affecting cardio-respiratory functions. *Frontiers in Aging Neuroscience, 5*, 5. doi:10.3389/fnagi.2013.00005

Keeler, J. R., Roth, E. A., Neuser, B. L., Spitsbergen, J. M., Waters, D. J. M., & Vianney, J.-M. (2015). The neurochemistry and social flow of singing: Bonding and oxytocin. *Frontiers in Human Neuroscience, 9.* 518. http://doi.org/10.3389/fnhum.2015.00518

Kilgour, A. R., Jakobson, L. S., & Cuddy, L. L. (2000). Music training and rate of presentation as mediators of text and song recall. *Memory & Cognition, 28*, 700–710.

Koelsch, S. (2010). Towards a neural basis of music-evoked emotions. *Trends in Cognitive Sciences.14*(3),131–137. doi:10.1016/j.tics.2010.01.002

Koelsch, S. (2014) Brain correlates of music-evoked emotions. *Nature Reviews Neuroscience, 15*(3):170–180. doi:10.1038/nrn3666

Kosfeld, M., Heinrichs, M., Zak, P. J., Fischbacher, U., & Fehr, E. (2005). Oxytocin increases trust in humans. *Nature, 435,* 673–676. doi:10.1038/nature03701

Krumhansl, C. L. (1997). An exploratory study of musical emotions and psychophysiology. *Canadian Journal of Experimental Psychology/Revue Canadienne de psychologie expérimentale, 51*(4), 336–353. doi:10.1037/1196-1961.51 4.336

Larson, E. B., Wang, L., Bowen, J. D., McCormick, W. C., Teri, L., Crane, P., & Kukull, W. (2006). Exercise is associated with reduced risk for incident dementia among persons 65 years of age and older. *Annals of Internal Medicine. 144,* 73–81.

Laukka, P. (2007). Uses of music and psychological well-being among the elderly. *Journal of Happiness Studies, 8,* 215. doi:10.1007/s10902-006-9024-3

Laurin, D., Verreault, R., Lindsay, J., MacPherson, K., & Rockwood, K. (2001). Physical activity and risk of cognitive impairment and dementia in elderly persons. *Archives of Neurology, 58,* 498–504.

Ledger, A., & Baker, F. (2007). An investigation of long-term effects of group music therapy on agitation levels of people with Alzheimer's disease. *Aging & Mental Health, 11*(3), 330–338. doi:10.1080/13607860600963406

Lin, Y., Chu, H., Yang, C. Y., Chen, C. H., Chen. S. G., Chang, H. J., . . . Chou, K. R. (2011). Effectiveness of group music intervention against agitated behavior in elderly persons with dementia. *International Journal of Geriatric Psychiatry, 26*(7), 670–678. doi:10.1002/gps.2580

Lundqvist, L. O., Carlsson, F., Hilmersson, P., & Juslin, P. N. (2008). Emotional responses to music: Experience, expression, and physiology. *Psychology of Music, 37*(1), 61–90. doi:10.1177/0305735607086048

Ma, W., & Thompson, W. F. (2015). Human emotions track changes in the acoustic environment. *Proceedings of the National Academy of Sciences, 112*(47), 14563–14568. doi:10.1073/pnas.1515087112

MacDonald, R. A. R. (2013). Music, health, and well-being: A review. *International Journal of Qualitative Studies on Health and Well-Being, 8.* doi:10.3402/qhw.v8i0.20635

MacDonald, R. A. R., Kreutz, G., & Mitchell, L. A. (2012). What is music health and well-being and why is it important. In R. A. R MacDonald, G. Kreutz, & L. A. Mitchell (Eds.), *Music, health and well-being.* Oxford: Oxford University Press.

Massaia, M., Reano, A., Luppi, C., Santagata, F., Marchetti, M., & Isaia, G. C. (2018). Receptive music interventions improve apathy and depression in elderly patients with dementia. *Geriatric Care, 4,* 1. doi:10.4081/gc.2018.7248

McDougald, A. (Producer) & Rossato-Bennett, M. (Producer and Director). (2014). *Alive Inside: A Story of Music and Memory* [Documentary]. United States: Projector Media. www.aliveinside.us

Meymandi, A. (2009). Music, medicine, healing, and the genome project. *Psychiatry, 6*(9), 43–45.

Moore, K. S. (2013). A systematic review on the neural effects of music on emotion regulation: Implications for music therapy practice. *Journal of Music Therapy, 50*(3), 198–242. doi:10.1093/jmt/50.3.198

Mortby, M. E., Maercker, A., & Forstmeier, S. (2012). Apathy: a separate syndrome from depression in dementia? A critical review. *Aging Clinical and Experimental Research, 24*(4), 305–316. doi: 10.3275/8105

Moussard, A., Bigand, E., Belleville, S., & Peretz, I. (2014). Music as a mnemonic to learn gesture sequences in normal aging and Alzheimer's disease. *Frontiers in Human Neuroscience, 8,* 294. doi:10.3389/fnhum.2014.00294

Narme, P., Clément, S., Ehrlé, N., Schiaratura, L., Vachez, S., Courtaigne, B., . . . Samson, S. (2014). Efficacy of musical interventions in dementia: Evidence from a randomized controlled trial. *Journal of Alzheimer's Disease, 38*(2), 359–369. doi:10.3233/JAD-130893.

Norberg, A., Melin, E., & Asplund, K. (2003) Reactions to music, touch and object presentation in the final stage of dementia: An exploratory study. *International Journal of Nursing Studies, 40*(5), 473–479. doi:10.1016/S0020-7489(03)00062-2

North, A. C., & Hargreaves, D. J. (1999). Music and Adolescent Identity. *Music Education Research, 1*(1), 75–92. doi:10.1080/1461380990010107

Olsen, K. N., & Stevens, C. J. (2013). Psychophysiological response to acoustic intensity change in a musical chord. *Journal of Psychophysiology, 27*(1), 16–26. doi:10.1027/0269-8803/a000082

Omar, R., Henley, S. M., Bartlett, J. W., Hailstone, J.C., Gordon, E., Sauter, D.A., . . . Warren, J. D. (2011). The structural neuroanatomy of music emotion recognition: evidence from frontotemporal lobar degeneration. *Neuroimage, 56*(3), 1814–1821. doi:10.1016/j.neuroimage.2011.03.002.

Osman, S. E., Tischler, V., & Schneider, J. (2016). "Singing for the brain": A qualitative study exploring the health and well-being benefits of singing for people with dementia and their carers. *Dementia (London), 15*(6), 1326–1339. doi:10.1177/1471301214556291

Overy, K., & Molnar-Szakacs, I. (2009). Being together in time: Musical experience and the mirror neuron system. *Music Perception: An Interdisciplinary Journal, 26*(5), 489–504. doi:10.1525/mp.2009.26.5.489

Park, D. C., & Reuter-Lorenz, P. (2009). The adaptive brain: Aging and neurocognitive scaffolding. *Annual Review of Psychology, 60,* 173–196.

Peterson, D. A., & Thaut, M. H. (2007). Music increases frontal EEG coherence during verbal learning. *Neuroscience Letters, 412,* 217–221.

Phillips-Silver, J. (2009). On the meaning of movement in music, development and the brain. *Contemporary Music Review, 28*(3), 293 314. doi:10.1080/07494460903404394

Prickett, C. A., & Moore, R. S. (1991). The use of music to aid memory of Alzheimer's patients. *Journal of Music Therapy, 28*(2), 101–110. doi:10.1093/jmt/28.2.101

Raglio, A., Bellandi, D., Baiardi, P., Gianotti, M., Ubezio, M.C., Zanacchi, E., . . . Stramba-Badiale M. (2015). Effect of active music therapy and individualized listening to music on dementia: A multicenter randomized controlled trial. *Journal of the American Geriatrics Society, 63*(8):1534–1539. doi:10.1111/jgs.13558

Raglio, A., Bellelli, G., Traficante, D., Gianotti, M., Ubezio, M.C., Villani, D., & Trabucchi, M. (2008). Efficacy of music therapy in the treatment of behavioral and psychiatric symptoms of dementia. *Alzheimer Disease & Associated Disorders, 22*(2), 158–162. doi:10.1097/WAD.0b013e3181630b6f.

Rickard, N. S., & McFerran, K. (2012). *Lifelong engagement with music: Benefits for mental health and well-being.* New York: Nova Science

Ridder, H. M., Stige, B., Qvale, L. G., & Gold, C. (2013). Individual music therapy for agitation in dementia: An exploratory randomized controlled trial. *Aging & mental health, 17*(6), 667–678.

Sakamoto, M., Ando, H., & Tsutou, A. (2013). Comparing the effects of different individualized music interventions for elderly individuals with severe dementia. *International Psychogeriatrics, 25*(5), 775–784. doi:10.1017/S1041610212002256

Salimpoor, V. N., Benovoy, M., Larcher, K., Dagher, A., & Zatorre, R. J. (2011). Anatomically distinct dopamine release during anticipation and experience of peak emotion to music. *Nature Neuroscience, 14*(2), 257–262. doi:10.1038/nn.2726

Samson, S., Dellacherie, D., & Platel, H. (2009). Emotional power of music in patients with memory disorders. *Annals of the New York Academy of Sciences, 1169*, 245–255. doi:10.1111/j.1749-6632.2009.04555.x

Särkämö, T. (2017). Music for the ageing brain: Cognitive, emotional, social, and neural benefits of musical leisure activities in stroke and dementia. *Dementia, 17*(6), 670–685. doi:10.1177/1471301217729237

Särkämö, T., Altenmüller, E., Rodríguez-Fornells, A., & Peretz, I. (2016). Editorial: Music, brain, and rehabilitation: Emerging therapeutic applications and potential neural mechanisms. *Frontiers in Human Neuroscience, 10*, 103. doi:10.3389/fnhum.2016.00103

Särkämö, T., Laitinen, S., Numminen, A., Kurki, M., Johnson, J. K., & Rantanen, P. (2016). Pattern of emotional benefits induced by regular singing and music listening in dementia. *Journal of the American Geriatrics Society, 64*(2), 439–440. doi:10.1111/jgs.13963.

Särkämö, T., Tervaniemi, M., Laitinen, S., Numminen, A., Kurki, M., Johnson, J. K., & Rantanen, P. (2014). Cognitive, emotional, and social benefits of regular musical activities in early dementia: Randomized controlled study. *Gerontologist. 54*, 634–650. doi:10.1093/geront/gnt100

Satoh, M., Ogawa, J., Tokita, T., Nakaguchi, N., Nakao, K., Kida, H., & Tomimoto, H. (2014). The Effects of Physical Exercise with Music on Cognitive Function of Elderly People: Mihama-Kiho Project. *PLoS ONE, 9*(4), e95230. doi: 10.1371/journal.pone.0095230

Serafine, M. L., Crowder, R. G., & Repp, B. H. (1984). Integration of melody and text in memory for songs. *Cognition, 16*(3), 285–303.

Sihvonen, A. J., Särkämö, T., Leo, V., Tervaniemi, M., Altenmüller, E., & Soinila S. (2017). Music-based interventions in neurological rehabilitation. *Lancet Neurology, 16*, 648e660.

Simmons-Stern, N. R., Budson, A. E., & Ally, B. A. (2010). Music as a memory enhancer in patients with Alzheimer's disease. *Neuropsychologia, 48*(10), 3164–3167. doi:10.1016/j.neuropsychologia.2010.04.033

Simmons-Stern, N. R., Deason, R. G., Brandler, B. J., Frustace, B. S., O'Connor, M. K., Ally, B. A., & Budson, A. E. (2012). Music-based memory enhancement in Alzheimer's disease: Promise and limitations. *Neuropsychologia, 50*(14), 3295–3303. doi:10.1016/j.neuropsychologia.2012.09.019

Shimizu, N. Umemura, T., Matsunaga, M., & Hirai, T. (2017). Effects of movement music therapy with a percussion instrument on physical and frontal lobe function in older adults with mild cognitive impairment: A randomized controlled trial. *Aging & Mental Health, 22*, 1–13. doi:10.1080/13607863.2017.1379048

Solé, C., Mercadal-Brotons, M., Galati, A., & De Castro, M. (2014). Effects of group music therapy on quality of life, affect, and participation in people with varying levels of dementia. *Journal of Music Therapy, 51*(1), 103–125. doi:10.1093/jmt/thu003.

Son, G. R., Therrien, B., & Whall, A. (2002). Implicit memory and familiarity among elders with dementia. *Journal of Nursing Scholarship, 34*(3), 263–267.

Tang, Q., Zhou, Y., Yang, S., Thomas, W. K. S., Smith, G. D., Yang, Z., . . . Chung, J. W. (2018). Effect of music intervention on apathy in nursing home residents with dementia. *Geriatric Nursing, 39*(4), 471–476. doi:10.1016/j.gerinurse.2018.02.003.

Tarr, B., Launay, J., & Dunbar, R. I. M. (2014). Music and social bonding: "Self-other" merging and neurohormonal mechanisms. *Frontiers in Psychology, 5*, 1096. doi:10.3389/fpsyg.2014.01096

Ten Brinke, L. F., Bolandzadeh, N., Nagamatsu, L. S., Hsu, C. L., Davis, J. C., Miran-Khan, K., & Liu-Ambrose, T. (2015). Aerobic exercise increases hippocampal volume in older women with probable mild cognitive impairment: A 6-month randomized controlled trial. *British Journal of Sports Medicine, 49*(4), 248–254. doi:10.1136/bjsports-2013-093184

Thaut, M. H., & Hoemberg, V. (Eds.). (2014). Handbook of neurologic music therapy. New York: Oxford University Press.

Thaut, M. H., Kenyon, G. P., Schauer, M. L., & McIntosh, G. C. (1999). The connection between rhythmicity and brain function. *IEEE Engineering in Medicine and Biology Magazine, 18*(2), 101–108. doi:10.1109/51.752991

Thompson, W. F., & Russo, F. A. (2004). The attribution of meaning and emotion to song lyrics. *Polskie Forum Psychologiczne, 9*, 51–62.

Thompson, W. F., Schellenberg, E. G., & Husain, G. (2001). Arousal, mood, and the Mozart effect. *Psychological Science, 12*(3), 248–251. doi:10.1111/1467-9280.00345

Thompson, W. F., & Schlaug, G. (2015). The healing power of music. *Scientific American Mind, 26*, 33–41.

Trehub, S. (2003). The developmental origins of musicality. *Nature Neuroscience, 6*, 669–673. doi:10.1038/nn1084

Trehub, S. E., Unyk, A. M., & Trainor, L. J. (1993). Maternal singing in cross-cultural perspective. *Infant Behavior and Development, 16*, 285–295. doi:10.1016/0163-6383(93)80036-8

Verghese, J., Lipton, R. B., Katz, M. J., Hall, C. B., Derby, C. A., Kuslansky, G., Ambrose, A.F., Sliwinski, M., & Buschke, H. (2003). Leisure activities and the risk of dementia in the elderly. *The New England Journal of Medicine, 348*, 2508–2516. doi:10.1056/NEJMoa022252

Wan, C. Y., & Schlaug, G. (2010). Music making as a tool for promoting brain plasticity across the life-span. *Neuroscientist, 16*, 566–577.

Warren, J. D., & Rohrer, J. D. (2009). Ravel's last illness: A unifying hypothesis, *Brain, 132*, e114. doi:10.1093/brain/awn245

# PART II

# IMPACT OF MUSIC
# ON COGNITION AND EMOTION
# IN PEOPLE WITH DEMENTIA

# 4

# Melody, Memory, and Engagement in Alzheimer's Disease

*Ashley D. Vanstone and Lola L. Cuddy*

Anecdotal accounts of responsiveness to music in persons with advanced Alzheimer's disease (AD) are striking—enough so that a documentary film on the topic, *Alive Inside*, garnered the Audience Award at the 2014 Sundance Film Festival. Following publication of a case study on memory for music in a person with advanced AD (Cuddy & Duffin, 2005), members of the public contacted us to share their stories of how music had (in most instances) been a positive force in the lives of their loved ones with AD. While such anecdotes do not permit inference regarding the prevalence of persons with AD who respond in meaningful ways to music, they do suggest that such responses are not uncommon and that, when they occur, they carry personal significance. Given that memory impairment is a hallmark symptom of AD, these anecdotes raised the question of whether memory for music might somehow be spared, relative to memory for other types of information, by the disease trajectory of AD.

Over the past decade, consensus has emerged that, although musical memory processes are not universally immune to AD, some aspects of memory for music are more durable than others (Baird & Samson, 2009, 2015). Furthermore, there is accumulating evidence of preserved neurocognitive mechanisms that could facilitate musical functioning in AD (Jacobsen et al., 2015). As the music cognitive profile of AD becomes better understood, however, we are still left with the question of how preserved memory for music might be reflected in the daily musical lives of those living with AD. Do musical memories facilitate responding to music and taking pleasure in musical activities? In what ways might preserved musical memories add to a person's quality of life? Is the therapeutic potential of music constrained by the individual's memory for music?

While, as we argue below, it is a reasonable hypothesis that preserved musical memories play a role in the lived experience of music for persons with AD, this relationship remains to be studied empirically; it cannot be assumed that these preserved memories have a functional impact. By way of analogy, many individuals with congenital deficits in their ability to process musical pitch do demonstrate, nonetheless, some appreciation for music (McDonald & Stewart, 2008), and conversely, there are some listeners who find relatively little pleasure in music, which is not explained in terms of music processing deficits or anhedonia in other domains (Martínez-Molina, Mas-Herrero, Rodríguez-Fornells, Zatorre, & Marco-Pallarés, 2016; Mas-Herrero, Zatorre, Rodriguez-Fornells, & Marco-Pallarés, 2014). If those with congenital deficits in the ability to process musical sound may nonetheless enjoy hearing music, then perhaps the mechanisms underlying engagement with music in listeners with AD are, similarly, independent of memory for music.

There are a number of conceptual and methodological challenges to studying this question empirically, to evaluating the complexity of naturalistic musical behavior against the carefully controlled neuropsychological study of memory for music. This chapter proposes a path to overcoming these challenges, by developing the construct of *music engagement*. We offer a definition of what it means to engage with music, placing this construct within a broader model for organizing empirical findings and, ultimately, for informing music-based interventions in AD or other dementias. As an illustration of how this model might guide empirical research, we discuss preserved memory for long-familiar music, which we argue holds particular promise for explaining how music engagement occurs in AD.

## Describing Musical Life in Dementia: Methodological and Conceptual Challenges

To understand the functional implications of preserved music cognitive abilities, we must first define and measure what it means to respond to music or to function in musical activities. The nature of AD complicates this task in three ways. First, as a progressive disorder, AD results in ongoing cognitive and functional decline in multiple domains; neuropsychological capabilities that were fairly stable in adulthood become subject to variable rates of change with disease progression (Stopford, Snowden, Thompson, & Neary, 2008; Weintraub, Wicklund, & Salmon, 2012). As a result, a meaningful characterization of day-to-day musical functioning in AD must be grounded within a timeframe that makes sense with respect to ongoing decline in cognitive ability.

Second, AD-related cognitive deficits give rise to functional limitations in the activities of daily living—in fact, the presence of such limitations is a diagnostic criterion (American Psychiatric Association, 2013). These functional limitations, even if not specifically music-related, could impinge on the ability to engage in a preferred music activity. For example, listening to a favorite recording becomes a more complicated endeavor if one has forgotten how to operate the stereo system. AD-related functional impairment is a potential confound to understanding the specific musical aspects of a person's functioning. Activities that define the musical lives of many adults may become inaccessible for persons with AD, due to factors completely unrelated to their interest in music or ability to process musical stimuli.

Finally, impairment of memory—and later in the course of the disease, language—gives rise to additional methodological challenges in that they render some forms of self-report unreliable for AD participants. Individuals with AD are able to report more meaningfully regarding immediate affective experience (e.g., mood states) than about behavior and daily activities (e.g., current activities of daily living; Salmon et al., 2008). In clinical settings, information on functioning in daily activities is typically provided by a caregiver or close family member. There are a variety of well-validated questionnaires to measure functional impairment as reported by a proxy respondent (Gelinas, Gauthier, McIntyre, & Gauthier, 1999). Relying on third-party information, then, is manageable from a psychometric point of view insofar as it reflects observable behavior.

However, evaluating musical functioning based only on observable behavior raises concerns about validity. Current conceptions of musical experience in nonclinical populations encompass not only observable behavior, but also personal, subjective experience; most people, after all, describe their musical experience in both subjective and objective terms. There is the option of relying on behavioral observation and self-report with respect to the immediate experience of a certain musical stimulus or activity. Where feasible, physiological metrics such as galvanic skin response or heart rate variability may offer an additional window into participants' inner responses to music. These options yield data that reflect the specific context in which they were collected and may not accurately capture how music is entwined in participants' day-to-day lives. In our previous research (Vanstone, Wolf, Poon, & Cuddy, 2016), we have approached this dilemma with the assumption that inner experience will be reflected, at least to some extent, in a participant's affect and behavior. These are observable by those who know the participant well, even if the resulting characterization of subjective musical experience lacks the nuance that would be possible through self-report.

AD-related cognitive and functional impairment, even though not specific to music, constrains how we measure musical experience and may confound how we view the construct, at least if we define musical experience in AD exclusively through concepts that have emerged from research with nonclinical populations. Hypotheses concerning the real-world implications of preserved memory for music must account for dementia-specific factors while drawing upon extant knowledge concerning musical experience within the general adult population. In the following section, we develop the concept of *music engagement* as a conceptual bridge between musical functioning as evidenced in daily life and musical functioning as observed in tasks measuring memory for music.

## Music Engagement: A Conceptual Framework for Musical Experience in AD

Our working definition of *engagement* is borrowed from the Comprehensive Process Model (CPM), which proposes that engagement is "the act of being occupied or involved with an external stimulus" (Cohen-Mansfield et al., 2011). The CPM conceptualizes engagement by persons with dementia as resulting from joint and reciprocal influences of the *stimulus, person*, and *environment*; attributes of each are relevant to understanding instances when a person is engaged with a particular stimulus. This is a model based on observational data collected in dementia care environments and, as such, has a decidedly clinical orientation. Disengagement is endemic among persons living with dementia, and the CPM is a framework for identifying the conditions under which persons with dementia are more likely to be successfully engaged in meaningful activities; disengagement, in itself, is considered a worthwhile target for intervention.

As articulated by Cohen-Mansfield and colleagues, the CPM is generic with respect to stimulus type and to the specific ways in which a person engages, and their studies to date have examined engagement with a wide variety of stimuli, including music (Cohen-Mansfield, Marx, Dakheel-Ali, Regier, & Thein, 2010). However, the stimulus characteristics that they have studied are higher-order characteristics that apply across stimulus types. For example, personally relevant meaning is a stimulus attribute shown to promote engagement (Cohen-Mansfield, Marx, Thein, & Dakheel-Ali, 2010). A model stated in such broad terms, while well suited to addressing the clinical issue of engagement across multiple stimulus types, requires further elaboration if it is to offer a conceptual framework for linking findings on memory for music with research on clinical and functional outcomes.

We contend that the CPM is well suited to elaboration with specific regard to musical stimuli. Our musical lives are enacted in a wide variety of activities that, although differing significantly in their forms and complexity, can meaningfully be said to reflect "occupation or involvement" with musical stimuli. These activities or forms of engagement, are the manifestation of multiple underlying processes, including attention, perception, motor planning, affect, and memory. The extent to which these processes are active in any instance is contingent on the inherent demands of the specific musical situation, as well as on the capability of the person to enact these processes.

In the CPM, music engagement is considered to emerge from the joint and reciprocal influences of attributes belonging to the *person, stimulus*, and *environment*. These parameters include the person's capability to carry out the underlying processes necessary for a particular form of music engagement. In this sense, we consider memory for music (whether impaired or preserved) as an attribute of the person. Taken together, attributes of the person, stimulus, and environment enable the underlying processes that facilitate moment-to-moment instantiations of music engagement.

In the original statement of the CPM, engagement is seen as the endpoint, but in the case of music engagement, it is necessary to consider the further effects arising from these moments of engagement—the existence of such effects, after all, is central to the rationale for any music-based intervention. Extending the CPM, we consider the *outcomes* of engagement to encompass the effects of music engagement on the person and within the social environment. Placed within the CPM, music-therapeutic outcomes are seen as arising from the processes of engagement, which, in turn, are jointly and reciprocally determined by the person, the stimulus, and the environment.

In the following sections, we describe each component of the CPM for music engagement in more detail, highlighting the relevance of music engagement as a framework for integrating basic and applied research on music and dementia. We then turn to a specific discussion of memory for long-familiar music in AD, conceptualizing its functional and therapeutic implications in terms of the CPM. In closing, we explore the clinical potential created by conceptualizing music-based interventions as forms of music engagement.

## The Varied Forms of Music Engagement

Engagement, defined here as reflecting *occupation or involvement with musical stimuli*, is constructed in deliberately broad terms, which is necessary due to the many and varied ways that individuals engage with music. Music engagement is ubiquitous in the lives of most people, present both in music-focused activities, such as playing an instrument or attending concerts, and as an accompaniment to other activities. Listening to music is used to structure activities of daily living (de Nora, 2000), facilitate mood regulation (Saarikallio, 2008), foster social connection and identity (MacDonald, Hargreaves, & Miell, 2002), and make physical exercise more enjoyable (Karageorghis & Priest, 2011). All of these uses involve some level of occupation or involvement with a musical stimulus, even though the activities vary substantially in their content and form.

Furthermore, instances of music engagement that appear identical to the casual observer may, in fact, be carried out to different ends and, therefore, be activating different processes within the listener. For example, consider the act of sitting in a chair and listening attentively to a musical recording. One listener experiences a wave of nostalgia while recalling an old friend who enjoyed the same song, while another listens in preparation to write a critical review of the recording. The behavior of these two listeners is similar, and each listener is meaningfully occupied with the musical stimulus, but each is engaging through a distinct set of cognitive and affective processes. While at this point it would be premature to propose an exhaustive list of the processes through which engagement occurs, it is proposed that they exist on multiple levels, including the perceptual, motor, cognitive, affective, social/interpersonal.

The diversity in forms and processes of music engagement is particularly salient when conceptualizing the musical experiences of those living with dementia. With the progression of dementia, listeners typically lose the ability to enact some of the processes through which they previously engaged with music, while retaining other abilities relevant to music engagement. Even as this shifting profile of abilities affects the particular forms of music engagement that are accessible to the person with dementia, other forms of engagement (perhaps even with the same musical stimuli) remain possible. It is possible, for example, to enjoy hearing a piece even if one has forgotten its name.

## Characteristics of the Musical Person

If music engagement takes many forms, if its quantity and quality vary substantially across individuals, then we are left with the task of identifying which personal attributes contribute to these individual differences. From the perspective of the Comprehensive Process Model, engagement is shaped by attributes of the person, interacting with attributes of the stimulus and the environment; to understand the experience of music in dementia, we must understand the relevant characteristics of the person who is musically engaged. These attributes include those that characterized the person prior to the onset of dementia because, whether or not such characteristics are obscured by dementia-related impairment, they may persist in the memories of others, or they may continue to reside implicitly in the person's self-concept. (For a thoughtful theoretical discussion of the musical self in dementia, see Baird & Thompson, 2018.) For example, if a person has lost the ability to play a musical instrument, a personal attribute no longer apparent at the moment of engagement, that lost ability may hold continued significance if the person's awareness of that loss infuses certain music with sadness or frustration, or if the lost ability contributes to the caregiver's sense of estrangement from the person who once was.

The capacity to engage with music begins very early in life and develops across the lifespan. Infants are able to discriminate many of the tonal and rhythmic features fundamental to music (Trehub, 2003). Throughout childhood and adolescence, fundamental music perceptual abilities are shaped by culture, giving rise to more sophisticated processing of musical structure and emotion (Stalinski & Schellenberg, 2012). During adolescence, music acquires heightened importance in the development of emotion regulation, personal mastery, relationships, and identity (Laiho, 2004). Across the lifespan, musical experiences accumulate, giving musical inflection to autobiographical memories (Janata, Tomic, & Rakowski, 2007), the memories by which we construct ourselves (Conway & Pleydell-Pearce, 2000). Our musical histories, consequently, are rich with individual differences in the memories, expertise, and repertoires that come to define the musical part of ourselves. These histories contain the personal attributes that, whether in their absence or their persistence, interact with stimulus and environment to shape engagement.

Music education deserves special comment as an aspect of premorbid musical experience—an attribute of the person—with direct relevance to music engagement. Formal music training, given sufficient intensity and, presumably, concordance with the student's wishes and inclinations, shapes the student in ways that could be predicted to moderate the effects of dementia on music engagement. Formal music training affects the structure and function of the auditory system and other brain regions (Kraus & Chandrasekaran, 2010) as well as the organization of auditory processing in the central nervous system more broadly (Skoe & Kraus, 2012). Learning to read music and to play a musical instrument opens an individual to new forms of music activity. Later in life, sustained engagement in music activities, especially at an intensive level, may ameliorate the effects of cognitive aging (Hanna-Pladdy & Gajewski, 2012; Hanna-Pladdy & Mackay, 2011). While a positive correlation between music education and engagement has been observed in the general population (Vanstone et al., 2016), the extent to which this effect persists in dementia has not yet been empirically established (see Chapter 8 for a discussion of musicians with dementia). Given the neurological and functional mechanisms by which the engagement-education link could be mediated, music education background is, arguably, a personal attribute deserving consideration in relation to music engagement in dementia.

Formal music education is, of course, only one type of formative musical experience that may shape an individual's musical life. Musical culture is transmitted first within the family, later within peer and social groups, dictating the types of music with which an individual is likely to be familiar as well as the specific ways in which music is used in leisure activities or in marking important life events. In adolescence, peers come to influence the styles of music to which an individual gravitates, as membership in social groups is reinforced by shared musical interests (MacDonald et al., 2002).

Individual differences in personality also seem to influence some aspects of preferred musical style (Chamorro-Premuzic & Furnham, 2007; Dunn, de Ruyter, & Bouwhuis, 2012; Rentfrow, Goldberg, & Levitin, 2011).

In addition to the numerous personal attributes shaping music engagement across the lifespan, the progression of dementia brings with it a variety of cognitive and functional impairments, which complicate the person's capacity to engage with music; within the present model of music engagement, we also consider these as attributes of the person. There is substantial interindividual variation in the severity and types of dementia-related cognitive impairment. For AD, the prototypical neuropsychological presentation includes progressive anterograde amnesia with subsequent impairment of attention, language, visuospatial ability, and executive function (Weintraub et al., 2012), but variations in this pattern exist independent of disease severity (Ralph, Patterson, Graham, Dawson, & Hodges, 2003). While these impairments are not specific to music, there is evidence for the association of some similar deficits with amusia in stroke patients (Särkämö et al., 2009) and for variation in hedonic response to music between dementia syndromes (Fletcher et al., 2015; see Chapter 5). Similar to dementia-related functional impairments, individual patterns of neuropsychological deficits, quite apart from those specific to music cognition, shape the person's response to music as well as the ancillary functional abilities that allow for participation in music activities.

The personal attributes relevant to music engagement are acquired across the lifespan, in the form of memories, preferences, habits, and skills, which are then disrupted by the emergence of dementia-related cognitive and functional impairments. Arguably, the relationship between music cognitive abilities and engagement is best understood as occurring alongside the personal attributes that shape music engagement across the lifespan, as well as those that emerge with the development of dementia.

## Attributes of the Musical Stimulus

Engagement is more likely when the stimulus has personal relevance (Cohen-Mansfield, Thein, Dakheel-Ali, & Marx, 2010). Memories of a melody give it personal relevance by virtue of their extra-musical associations and broader cultural resonance. In this way, preserved memory for long-familiar melodies is a personal attribute that binds person and stimulus. The person can be said to possess these preserved memories (an attribute of the person), while the melody can be said to be one that is remembered by that person (an attribute of the stimulus). The familiarity of a melody, then, is a stimulus attribute likely to promote engagement.

Familiarity is only one attribute of a musical stimulus with potential relevance to engagement in dementia, although relative to other aspects of musical structure, it has received more attention in the literature to date and has been a focus of the present discussion. We might also examine aspects of rhythm and meter, tonal complexity, or instrumentation, to name a few examples. These could be manipulated experimentally in terms of the functional dissociation of melodic processing (c.f. Peretz & Coltheart, 2003), the design features of music proposed in relation to the musical self (c.f. Baird & Thompson, 2018), or neurocognitive models of music perception and emotion (c.f. Koelsch, 2011).

Further understanding the relationship between musical structure and engagement would illuminate the mechanisms by which music might elicits hedonic or affective responses and engagement-related behavior. In clinical settings, it could also serve as a useful heuristic for identifying the music most likely to be engaging for patients whose pattern of dementia-related neuropsychological impairment has caused changes to musical preferences (see Chapter 5).

Music cognition abilities, a collection of abilities relevant to processing and remembering music, exist at the intersection of person and stimulus characteristics. The familiarity of a song and its historical location within a person's musical culture, as well as its musical structure, are characteristic of the stimulus. These characteristics bind the song to the person by virtue of corresponding personal characteristics, such as spared or impaired music processing abilities or the person's specific musical memories and preferences.

## Attributes of the Musical Environment

In any given instance of music engagement, the person is encountering the stimulus within a specific environment; characteristics of the environment facilitate or constrain how the person is able to engage. This aspect of music engagement has received less attention in the literature on music and dementia, although generally speaking the presence of distractions and noise is known to detract from engagement with a stimulus (Cohen-Mansfield et al.,

2011). Understanding the optimal environmental conditions for music engagement has clinical implications insofar as aspects of the dementia care environment could be optimized so as to enable rather than hinder the person's ability to engage.

Dementia-related functional impairments make it less likely that the person is able to shape the environment independently. Within residential dementia care units, design features allowing for privacy, space to wander, and home-like social spaces are associated with positive outcomes on a variety of behavioral health outcomes (Zeisel et al., 2003). Substantial evidence points to environmental factors as an important component in managing behavioral and psychological symptoms of dementia (Kales, Gitlin, & Lyketsos, 2015).

Ambient noise is predicted to have a deleterious effect on engagement with music. Age-related sensory hearing loss and cognitive changes in the auditory system make following conversations more difficult, even for typically aging adults (Schneider, Daneman, Pichora-fuller, & Columbia, 2002). AD-related neuropsychological impairment is associated with further difficulty tracking sounds in environments where there are multiple competing sources of sound (Hardy et al., 2016). Excluding extraneous noise from dementia-care settings seems to be an obvious step towards optimizing the auditory environment for music engagement, although the relationship between environmental noise and music perception for listeners with dementia has not been the subject of direct empirical study.

Engaging with music often requires resources in the form of stereo equipment or musical instruments, and so the availability of such resources might be considered an attribute of the environment. Without proper means of producing a musical stimulus, the potential forms of music engagement are curtailed. While this point might seem self-evident, it deserves special consideration in light of the functional impairment experienced by persons living with dementia: even if such resources are physically present, they may not actually be available to the person to the extent that the person lacks the functional ability to access and use the resources. A seeming lack of interest in listening to recordings, for example, could stem from an inability to operate the stereo system, or if the equipment is not readily visible, from having forgotten that it is present. Impairments of motor skills or visual perception have the potential to prevent the person from autonomously accessing and enjoying music. Modifying the environment to increase the salience of music-related cues and to make music easier to access could increase rates of spontaneous music engagement, although to our knowledge this concept has not been implemented or studied in clinical settings.

Experimental methods such as those used in the study of music and memory require careful control of environmental conditions to avoid injecting unnecessary error into the data. Participation in a music psychology experiment is, in essence, a form of music engagement, albeit one that is constructed so as to factor out the influence of environmental variables on the participant's responses to musical stimuli. As a result, the experimental literature on memory for music in dementia is based on analysis of person and stimulus, while setting aside environmental conditions. This approach has yielded substantial findings concerning the cognitive and neuropsychological mechanisms of music engagement, leaving open the intriguing question of how these mechanisms are deployed under the variety of environmental conditions in which listeners with dementia find themselves.

## Outcomes of Engagement

Music engagement is construed here as reflecting a person's involvement with a musical stimulus, as reflected in the proximal behavior surrounding presentation of musical stimuli. Engagement is evidenced by a wide range of behavior, including modulation of affect, motor behavior, or verbalization. The experience of engagement may indeed have value for its own sake, but it is hoped that encountering music has further short-term or long-term effects. In terms of our model, the flow-on effects of music engagement are termed *outcomes*. These are taken to encompass all of the consequences arising from engagement, whether they arise in the course of a music-based clinical intervention or as a consequence of recreational uses of music.

It is clear that music engagement has significant implications for the lifestyle and well-being of listeners across the lifespan. Use of music is positively associated with well-being (Laukka, 2007; Weinberg & Joseph, 2016), and regular music activities provide meaningful cognitive, emotional, and social benefits to those living with dementia (Elliott & Gardner, 2016; Särkämö et al., 2014). Shared music making activities are enjoyable and promote interpersonal connectedness not only for persons with dementia but also for their family and professional caregivers (McDermott, Orrell, & Ridder, 2014). For example, where one member of a couple is living with dementia, group singing activities provide an enjoyable experience in which both members of the couple are able to actively participate, making a positive contribution to the relationship (Unadkat, Camic, & Vella-Burrows, 2016; see Chapter 13). Music engagement in AD is significant as an avenue for enhancing quality of life through positive and meaningful shared activity, a striking contribution given that the cognitive and functional impairments of dementia place so many other activities out of reach. All of these benefits can be considered as outcomes of music engagement.

Music therapy is a common feature of dementia care, one for which evidence of efficacy continues to accumulate (see Chapters 9–13). It could be considered a special case of music engagement, one where engagement is facilitated in a clinical context by a professional with one or more therapeutic objectives. Modes of music-based intervention include incorporating music into psychological or physical rehabilitation, group-based music activities for caregivers and persons with dementia, and facilitating individualized listening. Regardless of the form these interventions take, they have in common that the clinician is presenting musical stimuli under conditions designed to prompt the person to engage. While the person's participation in the intervention likely evidences music engagement (unless the person were responding only to nonmusical features of the situation), the effects of the intervention are considered in the CPM to be outcomes of engagement.

## From Music Cognition to Music Engagement

As described earlier, applying the CPM to music engagement was motivated in part by the desire to understand the personal and therapeutic implications of research findings regarding memory for music in AD. Within the CPM, we have conceptualized the status of a person's capacity to process and remember music as forming part of the personal attributes relevant to how they engage with music. By placing these music cognitive abilities within a larger conceptual framework, we hope to clarify the mechanisms by which they might influence day-to-day musical experience of those living with AD. This model could be applied to numerous aspects of music processing and memory, but in this section, we focus on memory for long-familiar melodies (MLFM)—one particular aspect of memory for music—in order to illustrate how the CPM might be applied to interpreting research findings.

Although exciting instances have been documented in which individuals with AD have learned new music (Baird, Umbach, & Thompson, 2017) or used music to learn associated information (Simmons-Stern, Budson, & Ally, 2010), MLFM stands out as form of musical memory that is relatively robust in the face of AD. Recognition of familiar melodies is often preserved in persons with AD, sometimes even in individuals with severe AD (Cuddy et al., 2012). Relative to robustly encoded memory for long-familiar melodies, explicit memory for newly presented melodies is more likely to be impaired (Bartlett, Halpern, & Dowling, 1995; Ménard & Belleville, 2009; Vanstone et al., 2012), as is the ability to recall the names of familiar melodies (Kerer et al., 2013). Of the forms of musical memory studied to date, then, MLFM appears particularly resistant to cognitive decline in AD, which makes it a natural starting place in the larger project of characterizing the relationship between music cognition and engagement.

The relative durability of MLFM would not, in isolation, be expected to facilitate engagement; to have a positive effect on engagement, retrieving these memories should carry some reward value, either intrinsically or by facilitating secondary rewarding experiences. By way of comparison, it is hard to imagine that being able to recite the telephone directory would enhance quality of life in the absence of a functioning telephone. Long-familiar melodies are not inert, abstract facts, though. Rather, MLFM is implicated in a number of affective, cognitive, and behavioral processes that give rise, directly or indirectly, to rewarding experiences of music.

First, listeners tend to have more intense and pleasant emotional responses to familiar music relative to unfamiliar music (Ali & Peynircioǧǧlu, 2010). To the extent that preserved memory enhances the feeling of familiarity, it would be expected to enhance emotional response. The sense of familiarity has particular significance in the context of dementia care, where individuals face the disorienting experience of interacting with an environment that may seem increasingly less familiar; consider, for example, those nursing home residents who are not able to remember, despite the frequent and pained assurances of loved ones, that the dementia care unit is their new place of residence, and so are left in the perpetual nightmare of being stranded away from the familiar comfort of home.

Second, music-evoked autobiographical memories are more likely to be elicited by familiar music (Janata et al., 2007). This phenomenon extends to listeners with AD (El Haj, Clément, Fasotti, & Allain, 2013; El Haj, Postal, & Allain, 2012), for whom it tends to be a positive experience (Cuddy, Sikka, Silveira, Bai, & Vanstone, 2017). Having retained MLFM, then, has potential to facilitate access to other personally significant forms of memory, which in turn could be beneficial to the person.

Third, long-familiar music is often part of a repertoire that is widely known within a culture, thus creating potential for social connection through shared musical experience. Individuals are able to participate more meaningfully in group singing, for example, if they remember the melodies being sung. The social function of music begins in infancy, where infant-directed singing helps facilitate the parent-infant bond (Edwards, 2011). Music has a function of social cohesion in the spheres of leisure, courtship, work, war, and worship (Levitin, 2009). It is not surprising, then, to find that for couples where one partner has dementia, participation in a group singing activity resulted in positive relationship outcomes (Unadkat et al., 2016; see Chapter 13).

We have discussed preserved MLFM as a personal attribute that has significance with respect to important aspects of musical experience; engagement with long-familiar melodies has the potential to be engaging through at least the three mechanisms outlined here. If preserved MLFM, or other personal attributes, are expected to facilitate engagement, though, they are proposed to do so by means of interactions with the stimulus and the environment, each of which may have identifiable attributes that support engagement.

We addressed this hypothesis in a case series study of 15 participants with AD, which is described in detail elsewhere (Vanstone, 2017). Participants completed tests of melodic processing ability, memory for long-familiar melodies, and general cognitive impairment, and they provided questionnaire data on music training background and current levels of engagement in music-related activities. As a group, AD participants showed levels of music engagement that were similar to those of control participants, although some evidence suggested that higher levels of cognitive impairment were associated with a shift toward less complex music engagement activities. A comparative analysis of cases revealed multiple patterns of music cognitive ability and demographic characteristics associated with high levels of music engagement. In particular, preserved memory for long-familiar melodies showed the clearest association with music engagement among cases where AD-related cognitive impairment was more advanced. In light of the CPM, this analysis offers preliminary support for the contention that the status of a person's musical memories are, in fact, relevant to understanding engagement, but that this relationship occurs in the context of historical and current characteristics of the person.

## Concluding Comment: Toward Integration
## of Research and Practice

Scholarly research on music and dementia approaches the topic from the perspectives of multiple disciplines and uses a wide range of methodologies, framing questions in terms ranging from the neuropsychological to the ethnographic. This raises the challenge of how to integrate research findings and apply them clinically. The CPM is presented here as a framework for synthesizing the literature, using the construct of engagement as a point of convergence between the basic mechanisms and applied outcomes.

We have demonstrated the conceptual utility of this model through a discussion of our own work on memory for melodies, which illustrates the complex relationship between a basic neuropsychological mechanism (preserved MLFM) and patterns of music engagement reported in our participants. Other cognitive or perceptual processes might also be construed as personal attributes that, if spared by the dementia disease process, allow the person to experience meaningful responses to music. Similarly, other aspects of musical structure, such as rhythm or harmony, might be considered as stimulus attributes that shape patterns of music engagement in persons with AD or other dementias.

Music-based interventions, as framed in terms of our model, involve creating conditions in which the person, stimulus, and environment attributes align to facilitate forms of engagement that give rise to clinical outcomes. The clinician selects musical stimuli and arranges the interpersonal and physical environment in which the stimuli are presented, such that forms of engagement are cultivated to achieve an intended clinical outcome. For example, in an intervention to promote physical movement and wakefulness, the clinician will likely choose music with a lively beat (an attribute of the stimulus) that invites engagement in the form of clapping, dancing, and emotional arousal, which combat the clinical problem of daytime sleeping and inactivity (an outcome).

The CPM is structured in a way that lends itself naturally to clinical application, insofar as it frames engagement as shaped by complex individual factors interacting with stimulus and environment. Moving forward, a more comprehensive taxonomy of the characteristics that facilitate music engagement in AD, along with clinically appropriate ways of assessing these characteristics, would be a resource for case formulation and treatment planning in music therapy. With the CPM as an organizing tool, a clinician might identify the relevant characteristics of the person and the forms of engagement most likely to remediate the presenting clinical problem, selecting stimuli and shaping the environment in such a way that allows the person to engage productively with the music. For example, if disengagement were explained in one case by impairment of ADLs, then perhaps the clinician would need only to facilitate access to the activity—for example, by adapting the user interface on a music player so it is simpler to operate. On the other hand, if music perceptual mechanisms were compromised, the clinician may need to select musical material with simpler tonal structure or clearer rhythmic patterns. At each stage in this process, clinical decisions could be made with reference to basic or applied research findings regarding the relationships between each component of the CPM. Such a formulation would contribute to delivering sensitive care informed by evidence. Conversely, the richness of clinical observation could more fully inform the questions and hypotheses concerning the fundamental mechanisms and processes through which people, even while suffering from dementia, continue to express themselves as musical.

## References

Ali, S. O., & Peynircioğğlu, Z. F. (2010). Intensity of emotions conveyed and elicited by familiar and unfamiliar music. *Music Perception, 27*, 177–132.

American Psychiatric Association. (2013). *Diagnostic and statistical manual of mental disorders* (5th ed.). Arlington, VA: American Psychiatric Association.

Baird, A., & Samson, S. (2009). Memory for music in Alzheimer's disease: Unforgettable? *Neuropsychology Review, 19*, 85–101. https://doi.org/10.1007/s11065-009-9085-2

Baird, A., & Samson, S. (2015). Music and dementia. *Progress in Brain Research, 217*, 207–235. https://doi.org/10.1016/bs.pbr.2014.11.028

Baird, A., & Thompson, W. F. (2018). The impact of music on the self in dementia. *Journal of Alzheimer's Disease, 61*, 827–841.

Baird, A., Umbach, H., & Thompson, W. F. (2017). A nonmusician with severe Alzheimer's dementia learns a new song. *Neurocase, 23*, 36–40. https://doi.org/10.1080/13554794.2017.1287278

Bartlett, J. C., Halpern, A. R., & Dowling, W. J. (1995). Recognition of familiar and unfamiliar melodies in normal aging and Alzheimer's disease. *Memory & Cognition, 23*, 531–546. https://doi.org/10.3758/BF03197255

Chamorro-Premuzic, T., & Furnham, A. (2007). Personality and music: Can traits explain how people use music in everyday life? *British Journal of Psychology, 98*, 175–185.

Cohen-Mansfield, J., Marx, M. S., Dakheel-Ali, M., Regier, N. G., & Thein, K. (2010). Can persons with dementia be engaged with stimuli? *American Journal of Geriatric Psychiatry, 18*, 351–362. https://doi.org/10.1097/JGP.0b013e3181c531fd

Cohen-Mansfield, J., Marx, M. S., Freedman, L. S., Murad, H., Regier, N. G., Thein, K., & Dakheel-Ali, M. (2011). The comprehensive process model of engagement.

*American Journal of Geriatric Psychiatry, 19*, 859–870. https://doi.org/10.1097/
JGP.0b013e318202bf5b

Cohen-Mansfield, J., Marx, M. S., Thein, K., & Dakheel-Ali, M. (2010). The impact of past and present preferences on stimulus engagement in nursing home residents with dementia. *Aging & Mental Health, 14*, 67–73. https://doi.org/10.1080/13607860902845574

Cohen-Mansfield, J., Thein, K., Dakheel-Ali, M., & Marx, M. S. (2010). The underlying meaning of stimuli: Impact on engagement of persons with dementia. *Psychiatry Research, 177*, 216–222. https://doi.org/10.1016/j.psychres.2009.02.010

Conway, M. A., & Pleydell-Pearce, C. W. (2000). The construction of autobiographical memories in the self-memory system. *Psychological Review, 107*, 261–288. https://doi.org/10.1037//0033-295X.

Cuddy, L. L., & Duffin, J. (2005). Music, memory, and Alzheimer's disease: Is music recognition spared in dementia, and how can it be assessed. *Medical Hypotheses, 64*, 229–235. https://doi.org/10.1016/j.mehy.2004.09.005

Cuddy, L. L., Duffin, J. M., Gill, S. S., Brown, C. L., Sikka, R., & Vanstone, A. D. (2012). Memory for melodies and lyrics in Alzheimer's disease. *Music Perception, 29*, 479–491.

Cuddy, L. L., Sikka, R., Silveira, K., Bai, S., & Vanstone, A. (2017). Music-evoked autobiographical memories (MEAMs) in Alzheimer disease: Evidence for a positivity effect. *Cogent Psychology, 4*, 1277578. https://doi.org/10.1080/23311908.2016.1277578

de Nora, T. (2000). *Music in Everyday Life*. New York: Cambridge University Press.

Dunn, P. G., de Ruyter, B., & Bouwhuis, D. G. (2012). Toward a better understanding of the relation between music preference, listening behavior, and personality. *Psychology of Music, 40*, 411–428.

Edwards, J. (Ed.). (2011). *Music Therapy and Parent-Infant Bonding*. New York: Oxford University Press.

El Haj, M., Clément, S., Fasotti, L., & Allain, P. (2013). Effects of music on autobiographical verbal narration in Alzheimer's disease. *Journal of Neurolinguistics, 26*, 691–700.

El Haj, M., Postal, V., & Allain, P. (2012). Music enhances autobiographical memory in mild Alzheimer's disease. *Educational Gerontology, 38*, 30–41. https://doi.org/10.1080/03601277.2010.515897

Elliott, M., & Gardner, P. (2016). The role of music in the lives of older adults with dementia ageing in place: A scoping review. *Dementia, 0*, 1–15. https://doi.org/10.1177/1471301216639424

Fletcher, P. D., Downey, L. E., Golden, H. L., Clark, C. N., Slattery, C. F., Paterson, R. W., . . . Warren, J. D. (2015). Auditory hedonic phenotypes in dementia: A behavioural and neuroanatomical analysis. *Cortex, 67*, 95–105. https://doi.org/10.1016/j.cortex.2015.03.021

Gelinas, I., Gauthier, L., McIntyre, M., & Gauthier, S. (1999). Development of a functional measure for persons with Alzheimer's disease: The disability assessment for dementia. *American Journal of Occupational Therapy, 53*, 471–481.

Hanna-Pladdy, B., & Gajewski, B. (2012). Recent and past musical activity predicts cognitive aging variability: direct comparison with general lifestyle activities. *Frontiers in Human Neuroscience, 6*, 1–11. https://doi.org/10.3389%2Ffnhum.2012.00198

Hanna-Pladdy, B., & Mackay, A. (2011). The relation between instrumental musical activity and cognitive aging. *Neuropsychology, 25*, 378–386. https://doi.org/10.1037/a0021895

Hardy, C. J. D., Marshall, C. R., Golden, H. L., Clark, C. N., Mummery, C. J., Griffiths, T. D., . . . Warren, J. D. (2016). Hearing and dementia. *Journal of Neurology*, 1–16. https://doi.org/10.1007/s00415-016-8208-y

Jacobsen, J.-H., Stelzer, J., Fritz, T. H., Chételat, G., La Joie, R., & Turner, R. (2015). Why musical memory can be preserved in advanced Alzheimer's disease. *Brain, 138*, 2438–2450.

Janata, P., Tomic, S. T., & Rakowski, S. K. (2007). Characterisation of music-evoked autobiographical memories. *Memory, 15*, 845–860. https://doi.org/10.1080/09658210701734593

Kales, H. C., Gitlin, L. N., & Lyketsos, C. G. (2015). Assessment and management of behavioral and psychological symptoms of dementia. *The BMJ, 350.* https://doi.org/10.1136/bmj.h369

Karageorghis, C. I., & Priest, D.-L. (2011). Music in the exercise domain: A review and synthesis (Part I). *International Review of Sport and Exercise Psychology, 5*, 44–66. https://doi.org/10.1080/1750984X.2011.631026

Kerer, M., Marksteiner, J., Hinterhuber, H., Mazzola, G., Kemmler, G., Bliem, H. R., & Weiss, E. M. (2013). Explicit (semantic) memory for music in patients with mild cognitive impairment and early-stage Alzheimer's disease. *Experimental Aging Research, 39*, 536–564.

Koelsch, S. (2011). Toward a neural basis of music perception—a review and updated model. *Frontiers in Psychology, 2*, 110. https://doi.org/10.3389/fpsyg.2011.00110

Kraus, N., & Chandrasekaran, B. (2010). Music training for the development of auditory skills. *Nature Reviews Neuroscience, 11*, 599–605. https://doi.org/10.1038/nrn2882

Laiho, S. (2004). The psychological functions of music in adolescence. *Nordic Journal of Music Therapy, 13*, 47–63. https://doi.org/10.1080/08098130409478097

Laukka, P. (2007). Uses of music and psychological well-being among the elderly. *Journal of Happiness Studies, 8*, 215–241. https://doi.org/10.1007/s10902-006-9024-3

Levitin, D. J. (2009). *The world in six songs: How the musical brain created human nature.* New York: Plume.

MacDonald, R. A., Hargreaves, D. J., & Miell, D. (2002). *Musical identities.* Oxford University Press.

Martínez-Molina, N., Mas-Herrero, E., Rodríguez-Fornells, A., Zatorre, R. J., & Marco-Pallarés, J. (2016). Neural correlates of specific musical anhedonia. *Proceedings of the National Academy of Sciences, 113*, E7337–E7345. https://doi.org/10.1073/pnas.1611211113

Mas-Herrero, E., Zatorre, R. J., Rodriguez-Fornells, A., & Marco-Pallarés, J. (2014). Dissociation between musical and monetary reward responses in specific musical anhedonia. *Current Biology, 24*, 699–704. https://doi.org/10.1016/j.cub.2014.01.068

McDermott, O., Orrell, M., & Ridder, H. M. (2014). The importance of music for people with dementia: The perspectives of people with dementia, family carers, staff and music therapists. *Aging & Mental Health, 18*, 706–716. https://doi.org/10.1080/13607863.2013.875124

McDonald, C., & Stewart, L. (2008). Uses and functions of music in congenital amusia. *Music Perception, 25*, 345–355. https://doi.org/10.1525/mp.2008.25.4.345

Ménard, M. C., & Belleville, S. (2009). Musical and verbal memory in Alzheimer's disease: A study of long-term and short-term memory. *Brain and Cognition, 71*, 38–45. https://doi.org/10.1016/j.bandc.2009.03.008

Peretz, I., & Coltheart, M. (2003). Modularity of music processing. *Nature Neuroscience*, *6*, 688–691. https://doi.org/10.1038/nn1083

Ralph, L., Patterson, K., Graham, N., Dawson, K., & Hodges, J. R. (2003). Homogeneity and heterogeneity in mild cognitive impairment and Alzheimer's disease: a cross-sectional and longitudinal study of 55 cases. *Brain*, *126*, 2350.

Rentfrow, P. J., Goldberg, L. R., & Levitin, D. J. (2011). The structure of musical preferences: A five-factor model. *Journal of Personality and Social Psychology*, *100*, 1139.

Saarikallio, S. H. (2008). Music in mood regulation: Initial scale development. *Musicae Scientiae*, *12*, 291–309.

Salmon, E., Perani, D., Collette, F., Feyers, D., Kalbe, E., Holthoff, V., ... Herholz, K. (2008). A comparison of unawareness in frontotemporal dementia and Alzheimer's disease. *Journal of Neurology, Neurosurgery & Psychiatry*, *79*, 176–179. https://doi.org/10.1136/jnnp.2007.122853

Särkämö, T., Tervaniemi, M., Laitinen, S., Numminen, A., Kurki, M., Johnson, J. K., & Rantanen, P. (2014). Cognitive, emotional, and social benefits of regular musical activities in early dementia: Randomized controlled study. *Gerontologist*, *54*, 634–650.

Särkämö, T., Tervaniemi, M., Soinila, S., Autti, T., Silvennoinen, H. M., Laine, M., & Hietanen, M. (2009). Cognitive deficits associated with acquired amusia after stroke: A neuropsychological follow-up study. *Neuropsychologia*, *47*, 2642–2651.

Schneider, B. A., Daneman, M., Pichora-Fuller, M. K., & Columbia, B. (2002). Listening in aging adults : From discourse comprehension to psychoacoustics. *Aging*, 139–152.

Simmons-Stern, N. R., Budson, A. E., & Ally, B. A. (2010). Music as a memory enhancer in patients with Alzheimer's disease. *Neuropsychologia*, *48*, 3164–3167. https://doi.org/10.1016/j.neuropsychologia.2010.04.033

Skoe, E., & Kraus, N. (2012). A little goes a long way: How the adult brain is shaped by musical training in childhood. *The Journal of Neuroscience*, *32*, 11507–11510.

Stalinski, S. M., & Schellenberg, E. G. (2012). Music cognition: A developmental perspective. *Topics in Cognitive Science*, *4*, 485–497.

Stopford, C. L., Snowden, J. S., Thompson, J. C., & Neary, D. (2008). Variability in cognitive presentation of Alzheimer's disease. *Cortex*, *44*, 185–195.

Trehub, S. E. (2003). The developmental origins of musicality. *Nature Neuroscience*, *6*, 669–673. https://doi.org/10.1038/nn1084

Unadkat, S., Camic, P. M., & Vella-Burrows, T. (2016). Understanding the experience of group singing for couples where one partner has a diagnosis of dementia. *Gerontologist*, *0*, 1–11. https://doi.org/10.1093/geront/gnv698

Vanstone, A. D. (2017). *Music engagement in Alzheimer disease* (Doctoral dissertation). Queen's University, Kingston, Ontario, Canada. Retrieved from https://qspace.library.queensu.ca/bitstream/handle/1974/22698/Vanstone_Ashley_D_201709_PhD.pdf?sequence=3.

Vanstone, A. D., Sikka, R., Tangness, L., Sham, R., Garcia, A., & Cuddy, L. L. (2012). Episodic and semantic memory for melodies in Alzheimer's disease. *Music Perception*, *29*, 501–507. https://doi.org/10.1525/mp.2012.29.5.501

Vanstone, A. D., Wolf, M., Poon, T., & Cuddy, L. L. (2016). Measuring engagement with music: Development of an informant-report questionnaire. *Aging & Mental Health*, *20*, 474–484. https://doi.org/10.1080/13607863.2015.1021750

Weinberg, M. K., & Joseph, D. (2016). If you're happy and you know it: Music engagement and subjective wellbeing. *Psychology of Music*, *0*, 1–11. https://doi.org/10.1177/0305735616659552

Weintraub, S., Wicklund, A. H., & Salmon, D. P. (2012). The neuropsychological profile of Alzheimer disease. *Cold Spring Harbor Perspectives in Medicine, 2*, a006171. https://doi.org/10.1101/cshperspect.a006171

Zeisel, J., Silverstein, N. M., Hyde, J., Levkoff, S., Lawton, M. P., & Holmes, W. (2003). Environmental correlates to behavioral health outcomes in Alzheimer's special care units. *Gerontologist, 43*, 697–711. https://doi org/10.1093/geront/43.5.697

# 5

# Music Cognition in Frontotemporal Dementia and Non-Alzheimer's Dementias

*Rohani Omar*

## Introduction

Music cognition in non-Alzheimer's dementias has received far less research attention than in Alzheimer's disease. Much of the work to date involves patients with frontotemporal dementia (FTD). FTD consists of a group of diseases characterized by focal frontal and temporal lobe atrophy, and it is the second commonest cause of dementia affecting individuals under the age of 65 years. Patients with FTD exhibit clinical features dominated by a variety of behavioral and largely nonverbal cognitive disorders. These can include altered emotional and social responses, agnosias (the inability to process sensory information in any modality, resulting in problems recognizing objects, faces and smells, for example) and impairment in the regulation of physiological drives (such as appetite and sleep). Collectively, the FTD syndromes constitute a significant clinical and social burden, resulting in loss of independence, social dysfunction and isolation. The incidence of FTD is around 3–4 cases per 100,000 person-years (Snowden et al., 2002).

As described in the preface, FTD comprises three canonical clinical syndromes: behavioral-variant FTD (bv-FTD), semantic dementia, and progressive nonfluent aphasia. Patients with bv-FTD present predominantly with progressive change in personality and social behavior associated with executive dysfunction. Semantic dementia presents largely with difficulties in verbal comprehension and loss of word knowledge, resulting in naming problems and speech that is fluent but poor in content. In contrast, patients with progressive nonfluent aphasia present with progressive impairment of speech and language output and effortful, hesitant speech in the context of relatively preserved verbal comprehension. Pathologically, bv-FTD is associated with frontal and anterior temporal lobe atrophy. In semantic dementia, there is typically bilateral temporal lobe atrophy which is often asymmetric. Progressive nonfluent aphasia is associated with striking asymmetric atrophy of the left cortical hemisphere.

This chapter aims to provide an overview of the music cognition abilities in patients with non-Alzheimer's dementia with a focus on FTD. It will also address how the study of music cognition in these patients has helped shape our understanding of how our brains process musical information, from emotion recognition to associative music knowledge.

## What is Meant by Music Cognition?

In this chapter, the term music cognition is used to describe a variety of cognitive processes involving music as the stimulus, which include music perception (e.g., the ability to distinguish pitch, timbre, rhythm and dynamics), emotion processing from music (e.g., the ability to identify the predominant emotion conveyed in a piece of music, such as happiness, sadness, anger, or fear), as well as associative music knowledge, which is knowledge about a piece of music (e.g., the name of a song). As a complex sound, music is processed in the ascending auditory pathway to the primary auditory complex in Heschl's gyrus and the auditory association area in the planum temporale. The analysis of music involves perceptual components including pitch, timbre, and temporal structure. Knowledge of music is multidimensional, involving abstract "objects" such as notes and compositions, the representation of emotions, physical sources such as instruments, and symbols in the form of musical notation.

There has been debate surrounding the nature of "meaning" in music, and each of the dimensions above could be considered to convey meaning beyond purely perceptual sound features. In neuropsychology, the terms "meaning" and "knowledge" generally refer to learned concepts and facts about the world, and therefore "music knowledge" in neuropsychological studies describes the "association" of music with meaning based on learned attributes (associative knowledge of music), such as the ability to recognize a familiar song or identify the instrument that plays a piece of music. The processing of emotions in music can also be considered in this framework of music knowledge (Peretz & Coltheart, 2003).

## Why Study Music Cognition in FTD?

The FTD syndromes, and in particular bv-FTD, are often difficult to diagnose in the early stages due to the insidious nature of progressive behavioral decline which can be confused with primary psychiatric illnesses, such as depression, mania, obsessive-compulsive disorder, and schizophrenia (Wylie et al., 2013), or other dementia conditions. One of the main reasons for studying music cognition in FTD, from a clinical viewpoint, is as a novel metric that might shed some light on the underlying mechanisms and potentially help facilitate early diagnosis. In other words, understanding how people with different types of dementia respond to and process music might aid with the development of novel assessment tools for early identification of dementia and the development of music based treatments.

Another reason for studying music cognition in the FTD population is that it offers a complementary perspective to much of the work that has been conducted through functional neuroimaging and electrophysiology with healthy subjects (Halpern & Zatorre, 1999; Satoh et al., 2006; Steinbeis & Koelsch, 2008) and patients with focal brain injury (Ayotte et al., 2000; Eustache et al., 1990; Mendez, 2001; Schuppert et al., 2000; Stewart et al., 2006). While naturally occurring brain lesions are usually focal and not frequently directed towards anatomical locations of interest for music processing, neurodegenerative diseases affect large-scale brain networks including the frontal and temporal lobes which are likely to be critical for social cognition and music emotion processing, as discussed below (Seeley et al., 2009; Zhou et al., 2010).

The study of music cognition in FTD has been driven by the unique opportunities offered by these group of diseases into the understanding of brain processes in nonverbal sensory domains, including music. Certain neuropsychological functions relevant to music processing are characteristically impaired in FTD, such as semantic memory (the long-term memory store of conceptual information, including word knowledge, and facts about the world) in semantic dementia, and complex social and emotional behavior in bv-FTD. It is reasonable to suggest that difficulty in abstracting information from nonverbal signals could be one aspect behind the impairment in social skills and interpersonal behavior in FTD.

There is also considerable clinical interest in the potential therapeutic uses of music in patients with dementia. At present we unfortunately have little understanding of the effectiveness of music-based treatments in the non-Alzheimer's population. Much of the work on music therapy in dementia has focused on Alzheimer's disease (see Part 3 of this volume). There is an urgent need for further research into the responses of people with non-AD dementias to music.

## Studying Associative Music Knowledge in FTD

Designing neuropsychological tests of music knowledge aimed at patients with dementia is challenging given the wide range of cognitive impairments. The tests should probe one specific music-related ability each, with minimal reliance on other cognitive functions such as verbal input in order to remove as many confounds as possible. Examples include the use of two or three alternative forced choice tasks requiring matching of musical sounds or symbols to a visual target such as a written word, or to another auditory target such as a musical excerpt.

One of the earliest studies exploring music knowledge in dementia using tasks that assessed different music abilities was reported in 1993, when Polk and Kertesz published the two cases: a patient with progressive nonfluent aphasia whose brain MRI showed diffuse left cortical atrophy, and another patient with posterior cortical atrophy (a subtype of Alzheimer's disease). It was striking that the progressive nonfluent aphasia patient retained the ability to whistle continuations of familiar tunes following an introduction played on the piano, thus demonstrating sparing of expressive music functions, in contrast to severely impaired language production. The opposite pattern was shown in the posterior cortical atrophy patient in which language skills were preserved but music production ability was deficient with impaired melody and rhythm organization. The authors suggest that this finding of double dissociation between language and music functions supports the existence of distinct independent cognitive systems based on left (language) and right (music) laterality models.

Omar et al. (2010) is an example of a more recent study utilizing novel and specific music tasks to explore the cognitive model of music knowledge in non-Alzheimer's dementia, specifically in a patient with semantic dementia and another with Dementia with Lewy bodies. Dementia with Lewy bodies is a neurodegenerative disorder associated with the build-up of Lewy bodies (abnormal alpha-synuclein deposits) in the brain, and presents with cortical and subcortical cognitive impairments, particularly visuospatial

and executive dysfunction, with core clinical features of fluctuating attention, visual hallucinations, and Parkinsonism. Musical perceptual abilities
were initially assessed using the Montreal Battery of Evaluation of Amusia
(MBEA), a widely used test with normative data from musically untrained
subjects (Peretz et al., 2003). The following categories were then examined:

(a) *compositions*, in which the subjects were required to determine
    whether two melodic fragments were derived from the same musical
    composition. In addition, the patient with semantic dementia, who was
    still performing music regularly, was asked to play pieces of music ei-
    ther from memory given the name of the piece or, given his significant
    language deficits, following a musical introduction to each piece.
(b) *superordinate knowledge of musical objects*, whereby subjects were
    required to select the era of famous orchestral pieces from Baroque,
    Classical, Romantic, and 20th Century eras. The pieces were all written
    for a prominent solo instrument (mainly concertos), and the subjects
    were also asked to select the composer and solo instrument for which
    it was written (this solo instrument was not present in the excerpt
    presented) from three-alternative forced choice written word arrays.
    Determination of the (unheard) solo instrument would indicate spe-
    cific knowledge of the composition in question.
(c) *musical instruments*, whereby subjects were required to identify
    instruments from pictures as well as sounds, either by naming or
    miming, and further probing was conducted using a cross-modal pro-
    cedure matching audio clips to pictures and written names.
(d) *musical symbols*, in which a musical analogue of the language-based
    Synonyms Test (Warrington et al., 1998) was designed, whereby
    subjects were asked to determine whether two musical symbols are
    "synonyms" or equivalent (for example, a D flat is equivalent to a C
    sharp, and two quaver rests are equivalent to a crochet rest), thus
    demonstrating understanding of the musical meaning of each notation.

The semantic dementia patient had relatively preserved knowledge of musical compositions and symbols, despite demonstrating profoundly impaired
verbal skills. The Dementia with Lewy bodies patient, in contrast, showed
preserved knowledge of compositions and instruments despite significant
perceptual deficits. These findings support a modular organization of music
knowledge with some degree of dissociation between knowledge of musical

objects (compositions) and symbols, versus musical sources (instruments) and emotions. Superordinate knowledge of musical style, such as composer and era, was shown to be less vulnerable to disease than item-specific knowledge about compositions. This is similar to findings in other modalities such as language whereby the ability to identify a category of an object (e.g., "animal") is more robust than specifically naming the object (e.g., "giraffe").

It is suggested therefore that there exists a relatively independent associative knowledge system for music which is equivalent to semantic memory systems in other cognitive domains. This evidence considered together with other studies postulates that anatomical substrates for semantic processing of music may lie posterior and dorsal to the temporal lobe structures implicated in nonmusical semantic processing, and might thus explain the relative sparing of semantic memory for music in semantic dementia. Other areas of musical knowledge have also been shown to be preserved in semantic dementia and other FTD syndromes. There is a case report of a semantic dementia patient who remained able to sing popular tunes following a piano introduction despite being severely aphasic, indicating the retained ability to recognize music and produce musical output (Hailstone et al., 2009). Cases of preserved ability to play a musical instrument in people with AD or bv-FTD have also been described (see Chapter 8).

These observations demonstrate that careful cognitive assessment of music abilities using specific tasks can help to identify preserved skills in people with different types of dementia, even in the context of severe impairments in other cognitive domains such as language, episodic memory, and social cognition. The interest in studying music knowledge in patients with semantic dementia partly stems from the fact that their loss of object knowledge extends beyond language difficulties to include impairment of object recognition in multiple sensory domains such as faces and visual objects, environmental sounds (Bozeat et al., 2000; Goll et al., 2010), odors (Luzzi et al., 2007; Rami et al., 2007), flavors (Piwnica-Worms et al., 2010), and touch (Coccia et al., 2004). Indeed, impairments in music knowledge have been demonstrated in semantic dementia using a variety of tests, including song title-matching and detection of pitch errors in familiar tunes (Johnson et al., 2011), and making familiarity decisions for famous versus novel melodies (Hsieh et al., 2011). Neuroanatomically, deficits in pitch error detection were associated with right anterior temporal lobe atrophy, whilst tune familiarity performance correlated with grey matter atrophy in right anterior temporal lobe, insula, amygdala, and orbitofrontal lobe areas.

It is evident from the variety of musical skills differentially lost or preserved in FTD, particularly in semantic dementia where most of these studies have focused, that further work is required in larger populations of non-AD patients to better understand patterns of musical ability in these groups. There is, however, a particular aspect of music knowledge which has received greater research attention in FTD, and that is the processing of emotions in music.

## Emotion Processing of Music in FTD

There is accumulating evidence that music emotion processing abilities may be differentially affected by different dementia diseases (Drapeau et al., 2009; Raglio and Gianelli, 2009). It is well established that patients with FTD are impaired in processing emotional information from facial expressions (Kessels et al., 2007; Fernandez-Duque & Black, 2005; Rosen et al., 2004; Snowden et al., 2001) and voices (Keane et al., 2002; Snowden et al., 2008), and this impairment extends to emotion processing in music as well. This is in contrast to the relative sparing of musical emotion processing in Alzheimer's disease (Drapeau et al., 2009; Hsieh et al., 2012), suggesting that impairment in music emotion processing is not universal of neurodegenerative dementia diseases but may be relatively specific to certain degenerative pathologies, notably those in the FTD spectrum. What this highlights is the importance of having some degree of selectivity when designing clinical interventions such as music-based therapies targeting specific disease populations.

## The Evolutionary Role of Music in Humans: Why Is It Important to Us?

Apart from its purely perceptual elements, music has the extraordinary ability to generate powerful emotional responses, which is why it is so highly valued across human cultures. Judgements on the emotion expressed or generated from music have been found to be consistent among members of a musical culture (Peretz et al., 1998). Musical emotion processing is thought to involve brain mechanisms that are partly shared with those that process other emotional stimuli, including limbic structures, orbitofrontal cortex, and insula. Indeed, listening to music activates brain circuitry associated with pleasure and reward (Blood and Zatore, 2001; Koelsch et al., 2006; Menon and Levitin, 2005), and the neurotransmitter dopamine has been shown to be released in the striatal system at peak emotional arousal in response to music (Salimpoor et al., 2011). In addition, deficits of musical emotion comprehension have been reported following focal damage to these structures (Gosselin et al., 2006; Griffiths et al., 2004).

Unlike emotion-laden animate stimuli such as human faces and voices, music is an abstract entity without obvious survival value. So why is it such a fundamental part of human evolution? Darwin in 1859 had a keen interest in understanding song as a biological feature, and suggested that music and singing evolved with communicative capabilities as far back as our prelingual ancestors. Music was likely to have been essential to their survival, serving roles such as attracting a mating partner and warning each other about approaching predators (Schulkin & Raglan, 2014). Variations in pitch, rhythm and timbre enabled the communication of specific messages. As humans evolved, they lived in larger social groups and the need arose for more complex emotional expression and manipulation. In the absence of language, musicality may have been the principal manner in which emotions would have been expressed and manipulated (Mithen, 2009).

Music clearly serves a social role across human cultures. The powerful effect of music on emotions and mood is one of the main reasons to produce and listen to music (Juslin & Vastfjall, 2008). Koelsch (2010) describes the social functions of music, which include making contact with others, engaging social cognition, promoting communication and cooperation, and thus leading to increased social cohesion within groups. This has a functional purpose of promoting happiness and well-being. All this supports the argument that processing musical signals has similarities with the processing of other kinds of complex social and emotional signals within the human species.

## Music Emotion Processing Deficits in the FTD Spectrum of Diseases

One of the earlier studies of music emotion processing in dementia involved a patient with FTD who complained of difficulty identifying environmental sounds. The patient was shown on psychoacoustic testing not only to have auditory agnosia for environmental sounds but also impaired recognition of famous tunes and musical instruments. When presented with prerecorded environmental sounds, which included categories such as animal, human, musical instrument, and tools/equipment, the subject failed to accurately name any of the stimuli. He also misidentified sounds as positive versus negative in valence, for example clapping as negative and a siren as positive. He was unable to name or recognize some of his favorite pieces of music from his own CD collection, and he demonstrated impaired emotion recognition of musical excerpts that were happy, sad, peaceful, or scary. Despite this, he was still able to emotionally respond and derive great pleasure from listening to music, based on his own description of how much he enjoyed listening to his favorite music and also from perceived affective response during testing (Matthews et al., 2009). This demonstrates that the ability to identify perceptual characteristics of music and even to recognize the emotional intent in music is not necessarily required in order to experience an emotional response from music. To explain this phenomenon, the authors propose that there may be connectivity earlier than previously thought between musical perceptual processing regions in the brain and emotional response centers.

The differences in emotion recognition of musical versus other types of stimuli was further explored in a study of people with bv-FTD and semantic dementia (Omar et al., 2011). The patients were asked to select the dominant emotion expressed by a particular musical excerpt, in comparison to the same emotions in human faces and nonverbal vocal sounds Unsurprisingly, both patient groups scored significantly lower in emotion recognition across all modalities (faces, voices, and music) compared to healthy controls, but their performance was markedly worse for recognition of negative emotions fear and anger in music. Interestingly, no significant difference in performance was found between bv-FTD and semantic dementia subgroups. This deficiency of musical emotion processing correlated with an extensive bilateral cerebral network including insula, anterior cingulate, orbitofrontal, medial and dorsal prefrontal, and inferior frontal cortices, anterior and superior temporal cortices, fusiform and parahippocampal gyri, limbic structures, and other subcortical structures including nucleus accumbens and ventral tegmentum. Similar findings of music emotion recognition deficits were reported in other studies of FTD spectrum patients (Agustus et al., 2015; Hsieh et al., 2012; Macoir et al., 2016).

It was also noted in the Omar et al. study that music emotion recognition performance was associated with a number of brain areas not involved in facial or vocal emotion recognition. These "music-associated" areas are known to play a role in evaluating the mental states of others. Also known as "mentalizing," this cognitive ability to interpret the behavior of oneself and others in terms of mental states is characteristically impaired in bv-FTD. Downey and colleagues (2013) investigated mentalizing ability in a cohort of bv-FTD patients using music. In comparison to healthy controls, bv-FTD patients demonstrated impairment in the ability to attribute mental states but not nonmental characteristics to music. What these studies suggest is the possibility that music can represent surrogate mental states, in which the meaning of a piece of music is the emotional message it conveys, which requires active decoding by the brain based partly on learned associations from previous exposure to a musical culture or a particular piece of music. The neuroanatomical findings here provide circumstantial evidence for involvement of theory of mind processing in interpreting musical emotions, possibly in attributing a "mental state" to this abstract stimulus, as shown in fMRI evidence in healthy individuals (Steinbeis & Koelsch, 2009).

In FTD, this work shows a truly panmodal deficit of emotion understanding implicating not only animate emotional modalities such as faces and vocal expressions, but also "inanimate" abstract emotional stimuli such as music. From a neurobiological perspective, the cerebral areas associated with music emotion recognition include those previously shown to be implicated in processing emotional valence and intensity, "reward," coupling of subjective feeling states, and autonomic responses and representation of stimulus value (Adolphs et al., 1994; Blood & Zatorre, 2001; Calder et al., 2001; Cardinal et al., 2002; Dolan, 2007; Mitterschiffthaler et al., 2007).

In addition to the emotional processing deficits that have been observed in FTD, a phenomenon of enhanced music interest termed "musicophilia" has also been identified.

## Musicophilia in Bv-FTD

Abnormally enhanced reward-seeking behavior is another characteristic of patients with bv-FTD. Typically this involves stimuli with clear biological reward potential such as sweet food and drugs of abuse, although it is not restricted to these. Musicophilia, the abnormally intense craving for music, is also commonly reported in bv-FTD but is a poorly understood phenomenon. Fletcher et al. (2013, 2015) studied musicophilia in cohorts of dementia patients including bv-FTD, semantic dementia, progressive nonfluent aphasia, and AD. The studies identified that musicophilia was most common in people with bv-FTD and semantic dementia, but was not present in people with AD. Using voxel-based morphometry analysis, the presence of musicophilia was associated with increased grey matter volume in the left posterior hippocampus and cortical atrophy in posterior parietal and orbitofrontal areas, suggesting a possible brain substrate for musicophilia and the shifting of reward-seeking hedonic processing toward more abstract stimuli.

## Conclusions

There is a burgeoning body of work addressing various aspects of music ability in different dementia syndromes, and the results suggest a profile of music knowledge which varies not only between disease types but also between individuals, as described in case studies. The study of music knowledge in neurodegenerative dementias offers valuable neurobiological and clinical insights into brain mechanisms underpinning musical functions, which complement studies on patients with focal brain lesions. FTD patients in particular are known to have cognitive processing deficits in a variety of nonverbal modalities, and while music is an abstract entity with no obvious biological role in human survival, it clearly serves an important social role across cultures. Systematic experiments probing different areas of music knowledge have led to suggestion of a modular organization with some degree of dissociation. Anatomically the use of VBM and fMRI have allowed identification of brain substrates for certain musical functions. The demonstration of dissociated preservation of some forms of musical memory implies that music has the potential to access certain kinds of stored knowledge that might otherwise be impenetrable.

Core deficits in emotion processing in FTD may contribute to alterations of more complex social behaviors in these patients. It is suggested that the processing of emotion in music may act as a model system for the abstraction of emotions in complex real-life social situations and for the breakdown of emotional understanding in particular diseases. This would be consistent with the observation that, in contrast to FTD, comprehension of social signals and music emotion is often retained in Alzheimer's disease (Drapeau et al., 2009).

Future work should include the investigation of music emotion processing in other neurodegenerative conditions with impaired emotion encoding, such as Huntington's disease. It is evident that the effect on musical functions varies depending on the dementia type and cause. This means that patients with AD and non-AD dementias such as bv-FTD and semantic dementia are likely to respond differently to music-based therapies, and that such therapies need to be stratified and directed towards particular dementia syndromes.

# References

Adolphs, R., Tranel, D., Damasio, H., & Damasio A. (1994, December 15). Impaired recognition of emotion in facial expressions following bilateral damage to the human amygdala. *Nature, 372*(6507), 669–672.

Agustus, J. L., Mahoney, C. J., Downey, L. E., Omar, R., Cohen, M., White, M. J., ... Warren, J. D. (2015, March). Functional MRI of music emotion processing in frontotemporal dementia. *Annals of the New York Academy of Sciences, 1337*, 232–240.

Ayotte, J., Peretz, I., Rousseau, I., Bard, C., & Bojanowski, M. (2000). Patterns of music agnosia associated with middle cerebral artery infarcts. *Brain, 123*(9), 1926–1938.

Blood, A. J., & Zatorre, R. J. (2001 Sep 25). Intensely pleasurable responses to music correlate with activity in brain regions implicated in reward and emotion. *Proceedings of the National Academy of Sciences of the United States of America, 98*(20), 11318–11823.

Bozeat, S., Lambon Ralph, M. A., Patterson, K., Garrard, P., & Hodges, J. R. (2000). Non-verbal semantic impairment in semantic dementia. *Neuropsychologia, 38*(9), 1207–1215.

Calder, A. J., Lawrence, A. D., & Young, A. W. (2001). Neuropsychology of fear and loathing. *Nature Reviews Neuroscience, 2*(5), 352–363.

Cardinal, R. N., Parkinson, J. A., Hall, J., & Everitt, B. J. (2002). Emotion and motivation: The role of the amygdala, ventral striatum, and prefrontal cortex [Review]. *Neuroscience and Biobehavioural Reviews, 26*(3), 321–352.

Coccia, M., Bartolini, M., Luzzi, S., Provinciali, L., & Ralph, M. A. (2004). Semantic memory is an amodal, dynamic system: Evidence from the interaction of naming and object use in semantic dementia. *Cognitive Neuropsychology, 21*(5), 513–527.

Dolan, R. J. (2007). The human amygdala and orbital prefrontal cortex in behavioral regulation [Review]. *Philosophical Transactions of the Royal Society of London B: Biological Sciences, 362*(1481), 787–799.

Downey, L. E., Blezat, A., Nicholas, J., Omar, R., Golden, H. L., Mahoney, C. J., ... Warren, J. D. (2013). Mentalising music in frontotemporal dementia. *Cortex, 49*(7), 1844–1855.

Drapeau, J., Gosselin, N., Gagnon, L., Peretz, I., & Lorrain, D. (2009). Emotional recognition from face, voice, and music in dementia of the Alzheimer type. *Annals of the New York Academy of Sciences, 1169*, 342–345.

Eustache, F., Lechevalier, B., Viader, F., & Lambert, J. (1990). Identification and discrimination disorders in auditory perception: A report on two cases. *Neuropsychologia, 28*(3), 257–270.

Fernandez-Duque, D., & Black, S. E. (2005). Impaired recognition of negative fa-cial emotions in patients with frontotemporal dementia. *Neuropsychologia, 43*(11), 1673–1687.

Fletcher, P. D., Downey, L. E., Witoonpanich, P., & Warren, J. D. (2013). The brain basis of musicophilia: Evidence from frontotemporal lobar degeneration. *Frontiers in Psychology, 4,* 347.

Fletcher, P. D., Downey, L. E., Golden, H. L., Clark, C. N., Slattery, C. F., Paterson, R. W., . . . Warren, J. D. (2015). Auditory hedonic phenotypes in dementia: A behavioural and neuroanatomical analysis. *Cortex, 67,* 95–105.

Goll, J. C., Crutch, S. J., Loo, J. H., Rohrer, J. D., Frost, C., Bamiou, D. E., & Warren J. D. (2010). Non-verbal sound processing in the primary progressive aphasias. *Brain, 133*(Pt 1), 272–285.

Gosselin, N., Samson, S., Adolphs, R., Noulhiane, M., Roy, M., Hasboun, D., . . . Peretz, I. (2006). Emotional responses to unpleasant music correlates with damage to the parahippocampal cortex. *Brain, 129*(Pt 10), 2585–2592.

Griffiths, T. D., Warren, J. D., Dean, J. L., & Howard, D. (2004). "When the feeling's gone": A selective loss of musical emotion. *Journal of Neurology, Neurosurgery and Psychiatry, 75*(2), 344–345.

Hailstone, J. C., Omar, R., & Warren, J. D. (2009). Relatively preserved knowledge of music in semantic dementia. *Journal of Neurology, Neurosurgery and Psychiatry, 80*(7), 808–809.

Halpern, A. R., & Zatorre, R. J. (1999). When that tune runs through your head: A PET in-vestigation of auditory imagery for familiar melodies. *Cerebral Cortex, 9*(7), 697–704.

Hsieh, S., Hornberger, M., Piguet, O., & Hodges, J. R. (2011). Neural basis of music know-ledge: Evidence from the dementias. *Brain, 34*(Pt 9), 2523–2534.

Hsieh, S., Hornberger, M., Piguet, O., & Hodges, J. R. (2012). Brain correlates of musical and facial emotion recognition: Evidence from the dementias. *Neuropsychologia, 50*(8), 1814–1822.

Johnson, J. K., Chang, C. C., Brambati, S. M., Migliaccio, R., Gorno-Tempini, M. L., Miller, B. L., & Janata, P. (2011). Music recognition in frontotemporal lobar degeneration and Alzheimer disease. *Cognitive and Behavioral Neurology, 24*(2), 74–84.

Juslin, P. N., & Västfjäll, D. (2008). Emotional responses to music: The need to consider underlying mechanisms. *Behavioural Brain Science, 31*(5), 559–575.

Keane, J., Calder, A. J., Hodges, J. R., & Young, A. W. (2002). Face and emotion processing in frontal variant frontotemporal dementia. *Neuropsychologia, 40*(6), 655–665.

Kessels, R. P., Gerritsen, L., Montagne, B., Ackl, N., Diehl, J., & Danek, A. (2007). Recognition of facial expressions of different emotional intensities in patients with frontotemporal lobar degeneration. *Behavioural Neurology, 18*(1), 31–36.

Koelsch, S., Fritz, T. V., Cramon, D. Y., Müller, K., & Friederici, A. D. (2006). Investigating emotion with music: An fMRI study. *Human Brain Mapping, 27*(3), 239–250.

Koelsch, S. (2010). Towards a neural basis of music-evoked emotions. *Trends Cognitive Science, 14*(3), 131–137.

Luzzi, S., Snowden, J. S., Neary, D., Coccia, M., Provinciali, L., & Lambon Ralph, M. A. (2007). Distinct patterns of olfactory impairment in Alzheimer's disease, semantic de-mentia, frontotemporal dementia, and corticobasal degeneration. *Neuropsychologia, 45*(8), 1823–1831.

Macoir, J., Berubé-Lalancette, S., Wilson M. A., Laforce, R., Hudon, C., Gravel, P., . . . Monetta, L. (2016). When the Wedding March becomes sad: Semantic memory impairment for music in the semantic variant of primary progressive aphasia. *Neurocase, 22*(6), 486–495.

Matthews, B. R., Chang, C. C., De May, M., Engstrom, J., & Miller, B. L. (2009). Pleasurable emotional response to music, a case of neurodegenerative generalized auditory agnosia. *Neurocase, 15*(3), 248–259.

Mendez, M. F. (2001). Generalized auditory agnosia with spared music recognition in a left-hander. Analysis of a case with a right temporal stroke. *Cortex, 37*(1), 139–150.

Menon, V., & Levitin, D. J. (2005). The rewards of music listening: Response and physiological connectivity of the mesolimbic system. *Neuroimage, 28*(1), 175–184.

Mithen, S. (2009). The music instinct: The evolutionary basis of musicality. *Annals of the New York Academy of Sciences, 1169*(1), 3–12.

Mitterschiffthaler, M. T., Fu, C. H., Dalton, J. A., Andrew, C. M., & Williams, S C. (2007). A functional MRI study of happy and sad affective states induced by classical music. *Human Brain Mapping, 28*(11), 1150–1162.

Omar, R., Hailstone, J. C., Warren, J. E., Crutch, S. J., & Warren, J. D. (2010). The cognitive organization of music knowledge: A clinical analysis. *Brain, 133*(Pt. 4), 1200–1213.

Omar, R., Henley, S. M., Bartlett, J. W., Hailstone, J. C., Gordon, E., Sauter, D. A., . . Warren, J. D. (2011). The structural neuroanatomy of music emotion recognition: Evidence from frontotemporal lobar degeneration. *Neuroimage, 56*(3), 1814–1821.

Peretz, I., & Coltheart, M. (2003 Jul). Modularity of music processing. *Nature Neuroscience, 6*(7), 688–691.

Peretz, I., Gagnon, L., & Bouchard, B. (1998). Music and emotion: Perceptual determinants, immediacy, and isolation after brain damage. *Cognition, 68*(2), 111–141.

Piwnica-Worms, K. E., Omar, R., Hailstone, J. C., & Warren J. D. (2010). Flavour processing in semantic dementia. *Cortex, 46*(6), 761–768.

Raglio, A., & Gianelli, M. V. (2009). Music therapy for individuals with dementia: areas of interventions and research perspectives. *Current Alzheimer Research, 6*(3), 293–301.

Rami, L., Loy, C. T, Hailstone, J., & Warren, J. D. (2007). Odour identification in frontotemporal lobar degeneration. *Journal of Neurology, 254*(4), 431–435

Rosen, H. J., Pace-Savitsky, K., Perry, R. J., Kramer, J. H., Miller, B. L., & Levenson, R. W. (2004). Recognition of emotion in the frontal and temporal variants of frontotemporal dementia. *Dementia and Geriatric Cognitive Disorders, 17*(4), 277–281.

Salimpoor, V. N., Benovoy, M., Larcher, K., Dagher, A., & Zatorre, R. J. (2011). Anatomically distinct dopamine release during anticipation and experience of peak emotion to music. *Nature Neuroscience, 14*(2), 257–262.

Satoh, M., Takeda, K., Nagata, K., Shimosegawa, E., Kuzuhara, S. (2006). Positron-emission tomography of brain regions activated by recognition of familiar music. *American Journal of Neuroradiology, 27*(5), 1101–1106.

Schulkin, J., & Raglan, G. B. (2014). The evolution of music and human social capability. *Frontiers in Neuroscience, 17*(8), 1–13.

Schuppert, M., Münte, T. F., Wieringa, B. M., & Altenmüller, E. (2000). Receptive amusia: Evidence for cross-hemispheric neural networks underlying music processing strategies. *Brain, 123*(3), 546–559.

Seeley, W. W., Crawford, R. K., Zhou, J., Miller, B. L., & Greicius, M. D. (2009). Neurodegenerative diseases target large-scale human brain networks. *Neuron, 62*(1), 42–52.

Snowden, J. S., Neary, D., & Mann, D. M. (2002). Frontotemporal dementia. *British Journal of Psychiatry, 180*(2), 140–143.

Snowden, J. S., Austin, N. A., Sembi, S., Thompson, J. C., Craufurd, D., & Neary, D. (2008). Emotion recognition in Huntington's disease and frontotemporal dementia. *Neuropsychologia, 46*(11), 2638–2649.

Snowden, J. S., Bathgate, D., Varma, A., Blackshaw, A., Gibbons, Z. C., & Neary, D. (2001). Distinct behavioural profiles in frontotemporal dementia and semantic dementia. *Journal of Neurology, Neurosurgery and Psychiatry, 70*(3), 323–332.

Steinbeis, N., & Koelsch S. (2008). Shared neural resources between music and language indicate semantic processing of musical tension-resolution patterns. *Cerebral Cortex, 18*(5), 1169–1178.

Steinbeis, N., & Koelsch S. (2009). Understanding the intentions behind man-made products elicits neural activity in areas dedicated to mental state attribution. *Cerebral Cortex, 19*(3), 619–623.

Stewart, L., von Kriegstein, K., Warren, J. D., & Griffiths, T. D. (2006). Music and the brain: disorders of musical listening. *Brain, 129,* 2533–2553.

Warrington, E. K., McKenna, P., & Orpwood, L. (1998). Single word comprehension: A concrete and abstract word synonym test. *Neuropsychological Rehabilitation,* 8(2), 143–154.

Wylie, M. A., Shnall, A., Onyike, C. U., & Huey, E. D. (2013). Management of frontotemporal dementia in mental health and multidisciplinary settings. *International Review of Psychiatry, 25*(2), 230–236.

Zhou, J., Greicius, M. D., Gennatas, E. D., Growdon, M. E., Jang, J. Y., Rabinovici, G. D., . . . Seeley, W. W. (2010). Divergent network connectivity changes in behavioural variant frontotemporal dementia and Alzheimer's disease. *Brain, 133*(5), 1352–1367.

# 6

# Musical Leisure Activities to Support Cognitive and Emotional Functioning in Aging and Dementia

## A Review of Current Evidence

*Teppo Särkämö*

## Introduction

*"Music evokes emotion and emotion can bring with it memory."* This famous quotation from Dr. Oliver Sacks (1933–2015) elegantly summarizes one of the key facets of music, namely its unique ability to convey and elicit emotions and their close linkage to episodic and autobiographical memories. While this holds true for all of us, the emotional, communicative, and mnemonic impact of music is particularly evident in persons with dementia in whom musical activities are emerging as highly promising tools for supporting cognitive functioning, enhancing emotional well-being, and providing social interaction.

During the past decade, there has been increasing scientific interest towards the potential of music as a neurorehabilitation tool to facilitate functioning and well-being in severe neurological illnesses, including Alzheimer's disease (AD) and other types of dementia (for recent reviews, see Sihvonen et al., 2017; van der Steen et al., 2017). This interest has been fueled especially by advances in our understanding of the neural basis of music: functional neuroimaging studies in healthy participants have uncovered a wide-spread bilateral network of frontal, temporal, parietal, and limbic regions that govern many auditory, motor, cognitive, and emotional processes related to the perception and structural analysis (Koelsch & Siebel, 2005; Patel, 2003), production and expression (Beaty, 2015; Zatorre, Chen, & Penhune, 2007), enjoyment and reward (Koelsch, 2014; Zatorre &

Salimpoor, 2013), and memory (Janata, 2009; Janata, Tillmann, & Bharucha, 2002) of music. Importantly, many of these regions also show long-term structural neuroplastic changes as a result of regular musical training or activity (Brown, Zatorre, & Penhune, 2015; Herholz & Zatorre, 2012).

Given the aging of the population internationally (WHO, 2011), the rapidly increasing prevalence of dementia across the world (Prince et al., 2013) and the resulting growth of dementia-induced economic burden (Olesen et al., 2012), there is now an urgent need for cost-effective and easily applicable solutions to support the well-being and quality of life (QoL) of people with dementia and their family caregivers who provide the majority (70%) of primary care (Wimo & Prince, 2010) and who are often under great burden and psychological distress (Schneider, Murray, Banerjee, & Mann, 1999). Musical interventions have emerged as one of the most promising and viable nonpharmacological tools to meet this demand. Broadly speaking, music interventions can be classified either as *music therapy*, which is implemented by a trained music therapist and typically follows an established music-therapy protocol, or as *music medicine*, comprising various musical activities, such as music listening, which is implemented by other professionals (e.g., nursing staff), the patients themselves, or family caregivers. This chapter focuses on the latter category and provides an overview of the current evidence for the effects of musical leisure activities, such as music listening, singing, instrument playing, and dancing, on cognitive and emotional functioning in normal (healthy) aging as well as in dementia care and rehabilitation.

## Benefits of Musical Leisure Activities in Normal Aging

For most of us, music is present throughout the life-span, ranging from early childhood to old age, in one way or another and can serve a variety of functions and needs in everyday life, be they emotional, cognitive, motoric, or social in nature. Arguably, the key feature underlying all aspects of musical activity is the capacity for music to evoke and regulate emotions and to provide pleasure, comfort, and aesthetic fulfilment (Saarikallio, 2011; Zatorre & Salimpoor, 2013). Importantly, music-evoked emotions are mediated both by the autonomic nervous system (ANS) and the neuroendocrine (e.g., cortisol) system (Chanda & Levitin, 2013; Juslin & Västfjäll, 2008), indicating that the emotional impact of music has a deep-rooted neurobiological basis. In the next section, studies in healthy older adults exploring the impact of receptive musical activities (music listening) and expressive musical activities (instrument playing, singing, dancing) on emotional well-being and cognitive functioning are reviewed. It should be noted that for musical engagement, the distinction between the receptive and expressive activities is by no means an absolute one, as music listening may in some cases entail movement elements, such as spontaneous tapping or singing along.

## Receptive Musical Activities in Normal Aging

*Emotional and well-being effects.* Although the emotional and social impact of music in adolescence as a means for developing self-identity, social relationships, and emotion regulation skills is often emphasized, music continues to play an important role throughout adulthood and old age as a means of enhancing mood, evoking memories, and maintaining self-esteem and competence (Hays & Minichiello, 2005; Saarikallio, 2011). Socially, musical activities can also help in reducing loneliness and isolation, which is crucial given that low social participation, less frequent social contact, and more loneliness are all risk factors for the development of dementia (Kuiper et al., 2015). In older adults, music listening appears to be a common and important leisure activity that is linked to positive emotions and contributes to psychological well-being (Cohen, Bailey, & Nilsson, 2002; Kaufmann, Montross-Thomas, & Griser, 2018; Laukka, 2007). Using a large population-based sample (N = 5797), Kaufmann et al. (2018) found that 20% of older Americans did not listen to music whereas 75% reported average levels (between 1 and 28.5 hours per week) and 5% reported high levels of music listening. Interestingly, the average and high music listeners engaged more frequently also in other cognitive (e.g., playing games) and cultural activities (e.g., going to concerts or movies), social activities (e.g., visiting friends, communicating by telephone), and physical and ADL activities (e.g., walking, doing household chores) and also reported fewer somatic health conditions (e.g., lung disease, psychiatric problems) than the nonlisteners (Kaufmann et al., 2018). Another similar study of older U.S. adults (N = 1024) found that frequency of listening to religious music (especially gospel) was associated with lower self-reported death anxiety and higher life satisfaction, self-esteem, and sense of control (Bradshaw, Ellison, Fang, & Mueller, 2015).

Although these studies were cross-sectional and therefore do not provide any causal evidence for the effect of music, they nevertheless suggest that music listening is linked to a more enriched and healthy lifestyle and better mood in old age.

*Cognitive effects.* By inducing positive affect and heightened arousal, exposure to music (often with fast tempo and in major mode) can temporarily enhance cognitive performance (Thompson, Schellenberg, & Husain, 2001), also in elderly persons. Studies comparing the short-term effects of background music versus no music in older adults have reported an enhancement of performance on tasks of psychomotor speed (Bottiroli, Rosi, Russo, Vecchi, & Cavallini, 2014), verbal fluency (Thompson, Moulin, Hayre, & Jones, 2005), and episodic memory (Bottiroli et al., 2014; Ferreri et al., 2014) induced by the music. In contrast, one study found no effects of pretask music listening on working memory performance when compared to an auditory control (white noise) condition (Borella et al., 2017), and another study reported that background music had a distractive effect on performance in an associative memory task (Reaves, Graham, Grahn, Rabannifard, & Duarte, 2016). All in all, music listening may incur short-term cognitive benefits in older adults, but more studies are warranted to verify this effect. The long-term effects of regular music listening on cognitive functioning in healthy older adults have not been explored in controlled experimental studies.

## Expressive Musical Activities in Normal Aging

*Emotional and well-being effects.* In addition to music listening, the impact of different activities involving active music making or musical engagement (i.e., instrument playing, dancing, singing) on emotional well-being have been investigated. In a nonrandomized study of 29 older adults, Seinfeld and colleagues compared 4 months of piano lessons and daily training to other leisure activities (e.g., physical exercise, computer and painting lessons) and found that self-reported fatigue decreased and QoL related to physical and psychological health increased more in the piano intervention group (Seinfeld, Figueroa, Ortiz-Gil, & Sanchez-Vives, 2013). A recent within-subjects study of more intensive (30 hours, 3 hours per day) piano training in older adults reported enhanced musical self-efficacy after the training, but no significant transfer was observed to general self-efficacy or to physiological measures of stress indexed by cortisol (Bugos, Kochar, & Maxfield, 2016).

In a two-arm randomized controlled trial (RCT) by Hars and colleagues (2014), 134 community-dwelling older adults who were at increased risk for falling received either a group-based music intervention involving different multi-task exercises executed to the rhythm of music or no treatment (wait list control group) for 6 months. Compared to controls, the music intervention group showed a reduction in self-reported anxiety (Hars et al., 2014). Similarly, in a dancing study, Kattenstroth, Kalisch, Holt, Tegenthoff, and Dinse (2013) investigated the efficacy of a 6-month weekly dance training program (Agilando™) specifically developed for the elderly with 35 healthy older adults in a two-arm RCT study (dancing vs. no treatment control). The dance training was found to improve subjective well-being and contentment in life (Kattenstroth et al., 2013).

Also, participatory group-based musical activities, especially choir singing, have received increasing interest as potential ways of maintaining health and psychological well-being in aging. Questionnaire and interview studies and also more recent experimental studies of healthy older adults participating in community choirs have linked choir singing to various psychosocial and health-related benefits, suggesting that regular choral singing can bring about enjoyment, cognitive stimulation, better physical and mental health, and increased social interaction. In a questionnaire study, Johnson et al. (2013) explored the relationship between self-perceived benefits of choir singing and QoL and depressive symptoms in 177 Finnish older adults who sang in community choirs. The older choir singers reported few symptoms of depression as well as high overall QoL and satisfaction with health, and there was a significant relationship between the self-perceived benefits of choral singing and better QoL in the psychological, social relationships, and environmental domain (Johnson et al., 2013). Recently, using case-control sample of older choir singers (N = 109) and matched older adults drawn from the general population (N = 307), Johnson, Louhivuori, and Siljander (2017) found that the higher QoL and better health satisfaction in older choir singers was seen even after accounting for sociodemographic factors and engagement in other hobbies. Similarly, in a nonrandomized longitudinal study, Cohen et al. (2006) compared 90 older adults participating in a 30-week choir program with 76 control older adults over a 12-month follow-up and observed that choir singing was associated with better self-rating of health and morale, less loneliness, and higher level of activity.

Recently, the long-term efficacy of community singing has also been evaluated in an RCT in the UK. In a pioneering study, Coulton and colleagues

(2015) followed a large group (N = 258) of healthy older adults, half of whom participated in a 3-month community singing intervention, for 6 months using measures of QoL, mood, and health utility. The singing intervention had a long-term positive effect on health-related QoL as well as a short-term positive effect on mental health-related QoL, anxiety, and depression (Coulton et al., 2015). Overall, singing was reported to be more cost-effective than usual activities for increasing quality-adjusted life years (Coulton et al., 2015). Further, qualitative analysis of the subjective experiences of the participants in the trial provided converging results that the singing groups led to better physical, psychological, social, and community well-being (Skingley, Martin, & Clift, 2016).

*Cognitive effects.* The cognitive impact of music making in older adults has been explored both by looking at the effects of musical training earlier in life and of musical activities that take place in senior years. Regarding the former, studies have reported that elderly persons who have had instrumental musical training earlier in life have faster behavioral performance and neural timing in language tasks (Bidelman & Alain, 2015) as well as enhanced auditory attention (Zendel & Alain, 2014), melodic perception (Moreno-Gómez et al., 2017), and executive function, including working memory and cognitive control (Amer, Kalender, Hasher, Trehub, & Wong, 2013; Hanna-Pladdy & MacKay, 2011) compared to musically nontrained elderly persons. Regarding the latter, instrumental musical training that takes place at old age has been observed to improve cognitive performance in various tasks in healthy seniors (age ≥ 60 years). Bugos and colleagues (2007) randomized 31 older adults who had no previous musical training to a music intervention group receiving 6 months of individualized piano training or to a control group. Compared to the control group, the music group showed marked improvements on neurocognitive tests of attention, executive function, and processing speed, both immediately after the training period and in a 3-month longitudinal follow-up (Bugos et al., 2007). Similar results were also obtained in the piano training study of Seinfeld et al. (2013) in which 4 months of piano training enhanced performance on an executive test measuring inhibitory control and divided attention.

In addition to piano training, also the impact of other music-based interventions focusing on music and movement have been explored in healthy seniors. In the RCT conducted by Hars et al. (2014), 6 months of music-based multi-task training resulted in improvement in tests of general cognitive function (Mini-Mental State Examination, MMSE) and inhibitory control. Similarly, Maclean, Brown, and Astell (2014) explored the effects of a short rhythm-based training where older adults were trained to walk to the rhythm of music. Interestingly, the music training group performed better on a subsequent dual-task (performing a serial subtraction task while walking to the rhythm of music) than a group who had music playing in the background during the task but no training and a third control group who heard no music and received no training (Maclean et al., 2014). More widespread gains were observed in the dance training RCT by Kattenstroth et al. (2013): compared to the control group, the dance training group improved performance on a broad range of motor and cognitive functions, including posture, hand motor control, tactile processing, attention, memory, and processing speed. Importantly, the participants who benefited most from the intervention were those with lowest performance at baseline (Kattenstroth et al., 2013), suggesting that dancing could be applicable also to those elderly persons who are starting to show age-related cognitive or motor impairment. Another recent two-arm RCT compared a social dancing intervention (ballroom dancing, 8 months, twice weekly, 60 minutes per session) to a control physical intervention (home and community park walking) in 115 older adults and found a trend towards better postintervention performance in the dancing vs. walking group on a visuospatial memory task (Merom et al., 2016).

The association between musical activity and cognitive well-being has also been found in large-scale cross-sectional and cohort studies. In a landmark study, Verghese et al. (2003) assessed the relationship between different leisure activities and the later development of dementia over 5 years in 469 of persons aged above 75 years. Along with reading and playing board games, playing musical instruments and dancing were reported to be the leisure activities that were specifically linked to a reduced dementia risk (Verghese et al., 2003). More recently, using a cross-sectional design, Mansens, Deeg, and Comijs (2018) studied the association between self-reported singing and instrument playing activity and performance in neuropsychological tests in 1101 participants from the Longitudinal Aging Study Amsterdam aged ≥ 64 years. After controlling for potentially confounding variables, the older adults who reported playing an instrument scored higher on verbal learning, working memory, and processing speed than those not having any musical hobbies and also higher on processing speed than those who reported singing (Mansens et al., 2018). Regular singing, in contrast, has been linked to better preservation of voice production as indicated by more stable pitch and amplitude of the voice (Lortie, Rivard, Thibeault, & Tremblay, 2017).

## Benefits of Musical Leisure Activities in Dementia

Given the dramatically increasing prevalence of Alzheimer's disease (AD) and other dementia illnesses (Prince et al., 2013), there is a pressing need for effective ways to support cognitive, emotional, and social functioning in this population, both in people with dementia and their family members and caregivers. Importantly, music-induced emotions and memories are often preserved even in more advanced stages of dementia (Cuddy, Sikka, & Vanstone, 2015), possibly owing to relative preservation of medial frontal and limbic areas in AD (Jacobsen et al., 2015), which enables the therapeutic use of music across the dementia spectrum, from mild cognitive impairment (MCI) to severe dementia. Next, clinical studies exploring the impact of receptive and expressive musical leisure activities (excluding formal music therapy) on emotional well-being and cognitive functioning in dementia will be reviewed.

## Receptive Musical Activities in Dementia

*Emotional and well-being effects.* Studies exploring the emotional effects of receptive musical activities in dementia can be broadly divided between those assessing the immediate effect of music listening on mood state during single exposure and those assessing the short-term or long-term effects of more regular exposure to music. Regarding the former, pleasant and stimulating background music has been observed to temporarily reduce anxiety (Irish et al., 2006) and enhance awareness (Arroyo-Anlló, Díaz, & Gil, 2013). Regarding the latter, there are a number of small nonrandomized and randomized intervention studies in people with moderate-severe dementia residing in a long-term care facility that have assessed the emotional and social impact of musical leisure activities, primarily utilizing the listening of individualized (preferred) music, over a time period ranging from short (15–30 minutes) single sessions to multiple sessions over weeks. Using observational ratings, these studies have reported short-term beneficial effects of music listening on anxiety (Gerdner, 2005; Sung, Chang, & Lee, 2010), agitation (Garland, Beer, Eppingstall, & O'Connor, 2007; Remington, 2002; Ziv, Granot, Hai, Dassa, & Haimov, 2007), and positive social behaviors and interaction (Ziv et al., 2007).

Recently, there have also been a few RCT studies assessing the efficacy of interventions involving regular music listening. In a three-arm RCT by Raglio et al. (2015), music-therapy and music-listening interventions (10 weeks, twice a week, 30 minutes per session) were compared to standard care in 120 persons with moderate to severe dementia living in a nursing home, but no significant effects of the music interventions on behavioral and psychological symptoms of dementia (e.g., depression, anxiety) were observed. In another three-arm RCT, Cheung and colleagues (2018) compared active music-with-movement and music listening interventions to a social activity (chatting) intervention (all held for 6 weeks, twice a week, 30 minutes per session) in 165 people with dementia of moderate severity living in a nursing home and found that music listening enhanced memory performance in a delayed recall task compared to social activity. Innes et al. (2016) assessed the effects of two 12-week relaxation programs (daily practice, 12 minutes per day), one using Kirtan Kriya Meditation and one music listening, in 60 community-dwelling older adults with subjective cognitive decline (SCD) in a two-arm RCT. Both after the intervention period and in a 6-month follow-up, they observed significant improvements in psychological well-being, mood, and sleep quality in both groups, although these were slightly larger in the meditation than music listening group (Innes et al., 2016).

The studies listed above all used music listening interventions implemented by trained research assistants or rehabilitation personnel. Särkämö and colleagues (2014, 2016a, 2016b) recently performed a study exploring a novel dyadic *music coaching* intervention in which the caregivers (family members and nurses) were trained by a music therapist or singing teacher to use either music listening or singing together with the person with dementia as a part of daily care. In a three-arm RCT, dyads consisting of a person with mild-moderate dementia (half living at home, half at a nursing home) and their caregivers (N = 89) were randomized to two 10-week interventions, one utilizing singing and one music listening, and to a standard care control group. Both interventions comprised weekly group training sessions (90 minutes) as well as musical "homework" performed by the dyad at home between the sessions. The interventions entailed identifying which songs were emotionally and autobiographically most important to the person with dementia and instructing the caregivers on how to utilize them in everyday care for different purposes (e.g., relaxation, reminiscence, and vitalization). Outcome was assessed using neuropsychological tests as well as mood and QoL questionnaires performed at baseline, after the intervention, and 6 months later. Compared to standard care, music listening was found to be effective short-term (after the intervention period) in alleviating depressed mood, especially the behavioral signs of depression (agitation, loss of interest), and it also had a positive long-term effect on QoL seen at the 6-month follow-up (Särkämö et al., 2014, 2016b).

*Cognitive effects.* Studies assessing the immediate cognitive effects of short-term music exposure (background music) in persons with dementia have shown that hearing stimulating music can temporarily enhance cognitive performance in tasks of episodic (autobiographical) memory (El Haj, Fasotti, & Allain, 2012; Irish et al., 2006) and verbal fluency (Thompson et al., 2005). People with AD have also been shown to recall verbal material presented in a musical context (as song lyrics) better than material presented in a spoken context (Simmons-Stern, Budson, & Ally, 2010). A similar memory advantage of sung over spoken lyrics in immediate and/ or delayed recall has been observed also in two other AD studies (Moussard, Bigand, Belleville, & Peretz, 2014; Palisson et al., 2015). Together, these findings suggest that songs are a promising tool for accessing autobiographical memories and for the acquisition of novel verbal material in AD. To date, the long-term effects of regular music listening have been explored only in the RCT of Särkämö et al. (see above) where people living with dementia and caregivers listened to familiar songs together and discussed the emotions and memories they evoked. Compared to standard care, music listening had a positive short-term effect on general cognitive status (MMSE), attention, and executive function, as well as a long-term enhancing effect on orientation level and remote episodic memory (Särkämö et al., 2014). Interestingly, the beneficial cognitive effects of music listening were more evident in people with dementia of moderate than mild severity and non-AD-type dementia (Särkämö et al., 2016b).

## Expressive Musical Activities in Dementia

*Emotional and well-being effects.* Studies exploring the effects of expressive musical activities in dementia have mostly focused either on singing or on other group-based musical activities involving a combination of singing, moving to music, and instrument playing (e.g., accompanying songs with percussion instruments). In two RCTs, group-based live musical activities (4–8 weeks, 2–3 times per week, 40–60 minutes per session) were compared to a reading intervention in mild-moderate dementia (N = 47; Cooke, Moyle, Shum, Harrison, & Murfield, 2010) and to a cooking intervention in moderate-severe dementia (N = 48, Narme et al., 2014). The results indicated a general positive effect on emotional well-being (e.g., reduced depression and behavioral disorders, improved emotional state) across time, but no significant differences between the music and control interventions (Cooke et al., 2010; Narme et al., 2014). In contrast, the larger (N = 165) three-arm RCT of Cheung et al. (2018) in people with dementia of moderate severity found a music-with-movement intervention to be effective in reducing depression compared to a social activity (control condition).

The emotional and well-being effects of singing in dementia have thus far been explored in two RCTs and in one nonrandomized study. In the three-arm RCT of Särkämö et al. (see above), the singing intervention (familiar songs sung in a group and in a dyadic setting of a person with dementia and a caregiver at home) was found to be effective short-term in reducing depression, especially the physical signs of depression (e.g., lack of energy), as well as long-term in reducing caregiver stress and burden (Särkämö et al., 2014, 2016a). In a smaller non-RCT study on mild-moderate dementia (N = 20), Satoh et al. (2015) also observed that a 6-month (once a week, 60 minutes per session) karaoke-based singing training reduced neuropsychiatric symptoms and improved sleep time compared to a control group. Most recently, Pongan et al. (2017) assessed the effects of 12-week choir singing and painting interventions in 59 people with mild AD. Both interventions were reported to reduce anxiety and pain and improve QoL (no difference between the interventions here), whereas only the painting intervention reduced depression (Pongan et al., 2017). Further evidence concerning the social aspects of singing comes from qualitative studies of people with dementia and their family carers participating in group singing together. These studies have reported high engagement levels and positive effects on well-being (Camic, Williams, & Meeten, 2013) as well as improved dyadic and social relationships, identity and mood, and coping with dementia in daily life (Osman, Tischler, & Schneider, 2016; Unadkat, Camic, & Vella-Burrows, 2017).

*Cognitive effects.* The cognitive efficacy of expressive musical leisure activities in dementia has thus far been mapped in five RCTs (Cheung et al.,

2018; Doi et al., 2017; Narme et al., 2014; Pongan et al., 2017; Särkämö et al., 2014, 2016b) and two non-RCT studies (Maguire, Wanschura, Battaglia, Howell, & Flinn, 2015; Satoh et al. 2015). In the Narme et al. (2014) RCT, no effects of the music or cooking interventions on general cognitive status were observed, whereas in the Cheung et al. (2018) RCT an enhancement of memory storage and delayed recall performance was observed after the movement-to-music intervention compared to the social activity intervention. In the RCT of Särkämö and colleagues, compared to both standard care and to the music listening intervention, the singing intervention had a short-term positive effect on verbal working memory and it also maintained or improved general cognition, attention, executive function, and remote episodic memory (Särkämö et al., 2014). Importantly, the singing-induced enhancement of verbal working memory was seen particularly in persons with mild dementia, suggesting that the memory benefits of singing may be greater in the early stages of the illness. A similar finding was reported also in the Pongan et al. (2017) trial where a general improvement in working memory and executive function was seen in people with mild AD after both singing and painting interventions, but only the singing intervention had a positive impact on verbal memory. Maguire et al. (2015), in turn, found that a group singing intervention (4 months, 3 sessions per week) enhanced general cognition and visuospatial processing compared to a music listening intervention in people with dementia (N = 45). Satoh et al. (2015) found that psychomotor speed and mood were improved in people with dementia (N = 10) after a 6-month karaoke-based singing training program. Interestingly, these effects were coupled with decreased parietal activation in an fMRI karaoke task, suggestive of improved neural efficiency of cognitive processing (Satoh et al., 2015).

Finally, building on the key findings of the prospective cohort study of Verghese et al. (2003), Doi et al. (2017) recently reported a large-scale three-arm RCT of older adults with MCI (N = 201) that compared the cognitive effects of dancing and musical instrument playing interventions (40 weeks, once a week, 60 minutes per session) to a health education control intervention. The results of this pioneering study showed that both dancing and music playing improved general cognition (MMSE) and that dancing had an additional positive effect on verbal memory (Doi et al., 2017), suggesting that musical activities can serve as an effective means to combat age-related cognitive decline before the onset of dementia.

## Concluding Remarks

Based on the studies reviewed above, there is now emerging evidence that musical leisure activities or music-based interventions performed outside a formal music-therapy context can have many potential benefits for cognitive, emotional, and social functioning both in normal aging and in different stages of memory impairment. In summary, music listening has an enhancing effect on mood and arousal state, which can temporarily improve cognitive performance in attention or memory tasks in healthy older adults and persons with dementia, whereas evidence for the long-term beneficial effects of regular music listening is still rather scarce. Active musical hobbies, such as playing an instrument, singing, or dancing, have been shown to enhance executive functions, mood, or QoL in healthy older adults. As such, regular musical activities hold much promise as a way to maintain better mood and QoL and offset the gradual cognitive and neural decline associated with normal aging. There is also a potentially neuroprotective effect of music participation for neurodegenerative diseases, but long-term evidence (with many years of follow-up) for this is still lacking and more research is needed. In the dementia population, the specific impact of musical activities seems to depend on the severity of dementia symptoms: while positive effects on neuropsychiatric symptoms, such as agitation and social interaction, are seen in more advanced (severe) dementia, the cognitive benefits of music, particularly on memory (working memory, verbal memory), appear to be more pronounced in mild dementia. This suggests that the combination of cognitive, motor, and social stimulation provided by active music interventions could be important for slowing the progression of cognitive symptoms in the early stages of dementia, but, again, this claim needs to be tested in large-scale longitudinal trials with a follow-up spanning many years. The recent RCT results of Doi et al. (2017) on the positive effects of dancing and instrument playing on cognitive functioning over 10 months in elderly persons with MCI provide an important first step in this direction.

Overall, although positive findings from individual studies are converging to support the use of music in dementia care and rehabilitation, it must be noted these effects are not universal, as several of the studies covered here reported mixed effects on some affective and cognitive domains. Large and high-quality RCTs are still needed to build a more solid clinical evidence base and to establish the use of music more widely in rehabilitation and residential care units. Also studies looking at individual differences, musical features, and the mechanisms involved in response to music for older adults and persons living with dementia are required as well. Moreover, there is also a call for clinical music intervention studies combining behavioral outcome measures with neurophysiological and endocrinological markers, as well as structural and functional neuroimaging methods, that can better elucidate the neural mechanisms underlying the efficacy of music participation in dementia and, eventually, help target music interventions at the individual level to achieve maximal gains.

# References

Amer, T., Kalender, B., Hasher, L., Trehub, S. E., & Wong, Y. (2013). Do older professional musicians have cognitive advantages? *PLoS One, 8*, e71630.

Arroyo-Anlló, E. M., Díaz, J. P., & Gil, R. (2013). Familiar music as an enhancer of self-consciousness in patients with Alzheimer's disease. *BioMed Research International, 2013*, 752965.

Beaty, R. E. (2015). The neuroscience of musical improvisation. *Neuroscience & Biobehavioral Reviews, 51*, 108–117.

Bidelman, G. M. & Alain, C. (2015). Musical training orchestrates coordinated neuroplasticity in auditory brainstem and cortex to counteract age-related declines in categorical vowel perception. *Journal of Neuroscience, 35*, 1240–1249.

Borella, E., Carretti, B., Meneghetti, C., Carbone, E., Vincenzi, M., Madonna, J. C., . . . Mammarella, N. (2017). Is working memory training in older adults sensitive to music? *Psychological Research,* https://doi.org/10.1007/s00426-017-0961-8.

Bottiroli, S., Rosi, A., Russo, R., Vecchi, T., & Cavallini, E. (2014). The cognitive effects of listening to background music on older adults: Processing speed improves with up-beat music, while memory seems to benefit from both upbeat and downbeat music. *Frontiers in Aging Neuroscience, 6*, 284.

Bradshaw, M., Ellison, C. G., Fang, Q., & Mueller, C. (2015). Listening to religious music and mental health in later life. *Gerontologist, 55*, 961–971.

Brown, R. M., Zatorre, R. J., & Penhune, V. B. (2015). Expert music performance: Cognitive, neural, and developmental bases. *Progress in Brain Research, 217*, 57–86.

Bugos J. A., Kochar, S., & Maxfield, N. (2016). Intense piano training on self-efficacy and physiological stress in aging. *Psychology of Music, 44*, 611–624.

Bugos, J. A., Perlstein, W. M., McCrae, C. S., Brophy, T. S., & Bedenbaugh, P. H. (2007). Individualized piano instruction enhances executive functioning and working memory in older adults. *Aging & Mental Health, 11*, 464–471.

Camic, P. M., Williams, C. M., & Meeten, F. (2013). Does a "Singing Together Group" improve the quality of life of people with a dementia and their carers? A pilot evaluation study. *Dementia (London), 12*, 157–176.

Chanda, M. L. & Levitin, D. J. (2013). The neurochemistry of music. *Trends in Cognitive Science, 17*, 179–193.

Cheung, D. S., Lai, C. K., Wong, F. K., & Leung, M. C. (2018). The effects of the music-with-movement intervention on the cognitive functions of people with moderate dementia: a randomized controlled trial. *Aging & Mental Health, 22,* 306–315.

Cohen, A., Bailey, B., & Nilsson, T. (2002). The importance of music to seniors. *Psychomusicology, 18,* 89–102.

Cohen, G. D., Perlstein, S., Chapline, J., Kelly, J., Firth, K. M., & Simmens, S. (2006). The impact of professionally conducted cultural programmes on the physical health, mental health, and social functioning of older adults. *Gerontologist, 46,* 726–734.

Cooke, M., Moyle, W., Shum, D., Harrison, S., & Murfeld, J. (2010). A randomized controlled trial exploring the effect of music on quality of life and depression in older people with dementia. *Journal of Health Psychology, 15,* 765–776.

Coulton, S., Clift, S., Skingley, A., & Rodriguez, J. (2015). Effectiveness and cost-effectiveness of community singing on mental health-related quality of life of older people: Randomised controlled trial. *British Journal of Psychiatry, 207,* 250–255.

Cuddy, L. L., Sikka, R., & Vanstone, A. (2015). Preservation of musical memory and engagement in healthy aging and Alzheimer's disease. *Annals of the New York Academy of Sciences, 1337,* 223–231.

Doi, T., Verghese, J., Makizako, H., Tsutsumimoto, K., Hotta, R., Nakakubo, S., . . . Shimada, H. (2017). Effects of cognitive leisure activity on cognition in mild cognitive impairment: Results of a randomized controlled trial. *Journal of the American Medical Directors Association, 18,* 686–691.

El Haj, M., Fasotti, L., & Allain, P. (2012). The involuntary nature of music-evoked autobiographical memories in Alzheimer's disease. *Consciousness and Cognition, 21,* 238–246.

Ferreri, L., Bigand, E., Perrey, S., Muthalib, M., Bard, P., & Bugaiska, A. (2014). Less effort, better results: How does music act on prefrontal cortex in older adults during verbal encoding? An NIRS study. *Frontiers in Human Neuroscience, 8,* 301.

Garland, K., Beer, E., Eppingstall, B., & O'Connor, D. W. (2007). A comparison of two treatments of agitated behavior in nursing home residents with dementia: Simulated family presence and preferred music. *American Journal of Geriatric Psychiatry, 15,* 514–521.

Gerdner, L. A. (2005). Use of individualized music by trained staff and family: Translating research into practice. *Journal of Gerontological Nursing. 31,* 22–30.

Hanna-Pladdy, B. & MacKay, A. (2011). The relation between instrumental musical activity and cognitive aging. *Neuropsychology, 25,* 378–386.

Hars, M., Herrmann, F. R., Gold, G., Rizzoli, R., & Trombetti A. (2014). Effect of music-based multitask training on cognition and mood in older adults. *Age and Ageing, 43,* 196–200.

Hays, T., & Minichiello, V. (2005). The meaning of music in the lives of older people: A qualitative study. *Psychology of Music, 33,* 437–451.

Herholz, S. C., & Zatorre, R. J. (2012). Musical training as a framework for brain plasticity: Behavior, function, and structure. *Neuron, 76,* 486–502.

Innes, K. E., Selfe, T. K., Khalsa, D. S., & Kandati, S. (2016). Effects of meditation versus music listening on perceived stress, mood, sleep, and quality of life in adults with early memory loss: A pilot randomized controlled trial. *Journal of Alzheimer's Disease, 52,* 1277–1298.

Irish, M., Cunningham, C. J., Walsh, J. B., Coakley, D., Lawlor, B. A., & Robertson, I. H., & Coen, R. F. (2006). Investigating the enhancing effect of music on autobiographical memory in mild Alzheimer's disease. *Dementia and Geriatric Cognitive Disorders, 22,* 108–120.

Jacobsen, J. H., Stelzer, J., Fritz, T. H., Chételat, G., La Joie, R., & Turner, R. (2015). Why musical memory can be preserved in advanced Alzheimer's disease. *Brain, 138,* 2438–2450.

Janata, P. (2009). The neural architecture of music-evoked autobiographical memories. *Cerebral Cortex, 19,* 2579–2594.

Janata, P., Tillmann, B., & Bharucha, J. J. (2002). Listening to polyphonic music recruits domain-general attention and working memory circuits. *Cognitive, Affective, & Behavioral Neuroscience, 2,* 121–140.

Johnson, J. K., Louhivuori, J., & Siljander, E (2017). Comparison of well-being of older adult choir singers and the general population in Finland: A case-control study. *Musicae Scientiae, 21,* 178–194.

Johnson, J. K., Louhivuori, J., Stewart, A. L., Tolvanen, A., Ross, L., & Era, P. (2013). Quality of life (QOL) of older adult community choral singers in Finland. *International Psychogeriatrics, 25,* 1055–1064.

Juslin, P. N. & Västfjäll, D. (2008). Emotional responses to music: The need to consider underlying mechanisms. *Behavioral and Brain Sciences, 31,* 559–575.

Kattenstroth, J. C., Kalisch, T., Holt, S., Tegenthoff, M., & Dinse, H. R. (2013). Six months of dance intervention enhances postural, sensorimotor, and cognitive performance in elderly without affecting cardio-respiratory functions. *Frontiers in Aging Neuroscience, 5,* 5.

Kaufmann, C. N., Montross-Thomas, L. P., & Griser, S. (2018). Increased engagement with life: Differences in the cognitive, physical, social, and spiritual activities of older adult music listeners. *Gerontologist, 58,* 270–277.

Koelsch, S. (2014). Brain correlates of music-evoked emotions. *Nature Reviews Neuroscience, 15,* 70–80.

Koelsch, S., & Siebel, W. A. (2005). Towards a neural basis of music perception. *Trends in Cognitive Science, 9,* 578–584.

Kuiper, J. S., Zuidersma, M., Oude Voshaar, R. C., Zuidema, S. U., van den Heuvel, E. R., Stolk, R. P., & Smidt, N. (2015). Social relationships and risk of dementia: a systematic review and meta-analysis of longitudinal cohort studies. *Ageing Research Reviews, 22,* 39–57.

Laukka, P. (2007). Uses of music and psychological well-being among the elderly. *Journal of Happiness Studies, 8,* 215–241.

Lortie, C. L., Rivard, J., Thibeault, M., & Tremblay, P. (2017). The moderating effect of frequent singing on voice aging. *Journal of Voice, 31,* 112.

Maclean, L. M., Brown, L. J., & Astell, A. J. (2014). The effect of rhythmic musical training on healthy older adults' gait and cognitive function. *Gerontologist, 54,* 624–633.

Maguire, L. E., Wanschura, P. B., Battaglia, M. M., Howell, S. N., & Flinn, J. M. (2015). Participation in active singing leads to cognitive improvements in individuals with dementia. *Journal of the American Geriatrics Society, 63,* 815–816.

Mansens, D., Deeg, D. J., H., & Comijs, H. C. (2018). The association between singing and/or playing a musical instrument and cognitive functions in older adults. *Aging & Mental Health, 22,* 964–971.

Merom, D., Grunseit, A., Eramudugolla, R., Jefferis, B., Mcneill, J., & Anstey, K. J. (2016). Cognitive benefits of social dancing and walking in old age: The dancing mind randomized controlled trial. *Frontiers in Aging Neuroscience, 8*, 26.

Moreno-Gómez, F. N, Véliz, G, Rojas, M. Martínez, C., Olmedo, R., Panussis, F., . . . Delano, P. H. (2017). Music training and education slow the deterioration of music perception produced by presbycusis in the elderly. *Frontiers in Aging Neuroscience, 9*, 149.

Moussard, A., Bigand, E., Belleville, S., & Peretz, I. (2014). Learning sung lyrics aids retention in normal ageing and Alzheimer's disease. *Neuropsychological Rehabilitation, 24*, 894–917.

Narme, P., Clément, S., Ehrlé, N., Schiaratura, L., Vachez, S., Courtaigne, B., . . . Samson, S. (2014). Efficacy of musical interventions in dementia: Evidence from a randomized controlled trial. *Journal of Alzheimer's Disease, 38*, 359–369.

Olesen, J., Gustavsson, A., Svensson, M., Wittchen, H. U., Jönsson, B., & CDBE2010 study group (2012). The economic cost of brain disorders in Europe. *European Journal of Neurology, 19*, 155–162.

Osman, S. E., Tischler, V., & Schneider, J. (2016). "Singing for the Brain": A qualitative study exploring the health and well-being benefits of singing for people with dementia and their carers. *Dementia (London), 15*, 1326–1339.

Palisson, J., Roussel-Bacletm, C., Maillet, D., Belin, C., Ankri, J., & Narme, P. (2015). Music enhances verbal episodic memory in Alzheimer's disease. *Journal of Clinical and Experimental Neuropsychology, 37*, 503–517.

Patel, A. D. (2003). Language, music, syntax and the brain. *Nature Neuroscience, 6*, 674–681.

Pongan, E., Tillmann, B., Leveque, Y., Trombert, B., Getenet. J. C., . . . LACMé Group. (2017). Can musical or painting interventions improve chronic pain, mood, quality of life, and cognition in patients with mild Alzheimer's disease? Evidence from a randomized controlled trial. *Journal of Alzheimer's Disease, 50*, 663–677.

Prince, M., Bryce, R., Albanese, E., Wimo, A., Ribeiro, W., & Ferri, C. P. (2013). The global prevalence of dementia: A systematic review and meta-analysis. *Alzheimer's & Dementia, 9*, 63–75.

Raglio, A., Bellandi, D., Baiardi, P., Gianotti, M., Ubezio, M C., Zanacchi, E., . . . Stramba-Badiale, M. (2015). Effect of active music therapy and individualized listening to music on dementia: A multicenter randomized controlled trial. *Journal of the American Geriatrics Society, 63*, 1534–1539.

Reaves, S., Graham, B., Grahn, J., Rabannifard, P., & Duarte. A. (2016). Turn off the music! Music impairs visual associative memory performance in older adults. *Gerontologist, 56*, 569–577.

Remington, R. (2002). Calming music and hand massage with agitated elderly. *Nursing Research, 51*, 317–323.

Saarikallio, S. (2011). Music as emotional self-regulation throughout adulthood. *Psychology of Music, 39*, 307–327.

Särkämö, T., Laitinen, S., Numminen, A., Kurki, M., Johnson, J. K., & Rantanen, P. (2016a). Pattern of emotional benefits induced by regular singing and music listening in dementia. *Journal of the American Geriatrics Society, 64*, 439–440.

Särkämö, T., Laitinen, S., Numminen, A., Kurki, M., Johnson, J. K., & Rantanen, P. (2016b). Clinical and demographic factors associated with the cognitive and emotional efficacy of regular musical activities in dementia. *Journal of Alzheimer's Disease, 49*, 767–781.

Särkämö, T., Tervaniemi, M., Laitinen, S., Numminen, A., Kurki, M., Johnson, J. K., & Rantanen, P. (2014). Cognitive, emotional, and social benefits of regular musical activities in early dementia: randomized controlled study. *Gerontologist, 54,* 634–650.

Satoh, M., Yuba, T., Tabei, K., Okubo, Y., Kida, H., Sakuma, H., & Tomimoto, H. (2015). Music therapy using singing training improves psychomotor speed in patients with Alzheimer's disease: A neuropsychological and fMRI study. *Dementia and Geriatric Cognitive Disorders, 5,* 296–308.

Schneider, J., Murray, J., Banerjee, S., & Mann, A. (1999). EUROCARE: A cross-national study of co-resident spouse carers for people with Alzheimer's disease: I–Factors associated with carer burden. *International Journal of Geriatric Psychiatry, 14,* 651–661.

Seinfeld, S., Figueroa, H., Ortiz-Gil, J., & Sanchez-Vives, M. V. (2013). Effects of music learning and piano practice on cognitive function, mood and quality of life in older adults. *Frontiers in Psychology, 4,* 810.

Sihvonen, A. J., Särkämö, T., Leo, V., Tervaniemi, M., Altenmüller, E., & Soinila, S. (2017). Music-based interventions in neurological rehabilitation. *Lancet Neurology, 16,* 648–660.

Simmons-Stern, N. R., Budson, A. E., & Ally, B. A. (2010). Music as a memory enhancer in patients with Alzheimer's disease. *Neuropsychologia, 48,* 3164–3167.

Skingley, A., Martin, A., & Clift, S. (2016). The contribution of community singing groups to the well-being of older people: Participant perspectives from the United Kingdom. *Journal of Applied Gerontology, 35,* 1302–1324.

Sung, H. C., Chang, A. M., & Lee, W. L. (2010). A preferred music listening intervention to reduce anxiety in older adults with dementia in nursing homes. *Journal of Clinical Nursing, 19,* 1056–1064.

Thompson, R. G., Moulin, C. J., Hayre, S., & Jones, R. W. (2005). Music enhances category fluency in healthy older adults and Alzheimer's disease patients. *Experimental Aging Research, 31,* 91–99.

Thompson, W. F., Schellenberg, E. G., & Husain, G. (2001). Arousal, mood, and the Mozart effect. *Psychological Science, 12,* 248–251.

Unadkat, S., Camic, P. M., & Vella-Burrows, T. (2017). Understanding the experience of group singing for couples where one partner has a diagnosis of dementia. *Gerontologist, 57,* 469–478.

van der Steen, J. T., van Soest-Poortvliet. M. C., van der Wouden, J. C., Bruinsma, M. S., Scholten, R. J., & Vink, A. C. (2017). Music-based therapeutic interventions for people with dementia. *Cochrane Database of Systematic Reviews, 5,* CD003477.

Verghese, J., Lipton, R. B., Katz, M. J., Hall, C. B., Derby, C. A., Kuslansky, G., . . . Buschke H. (2003). Leisure activities and the risk of dementia in the elderly. *New England Journal of Medicine, 348,* 2508–2516.

Wimo, A., & Prince, M. (2010). *World Alzheimer Report 2010.* London: Alzheimer's Disease International.

World Health Organization (WHO). (2011). *Global Health and Ageing.* Geneva: World Health Organization.

Zatorre, R. J., Chen, J. L., & Penhune, V. B. (2007). When the brain plays music: Auditory-motor interactions in music perception and production. *Nature Reviews Neuroscience, 8,* 547–558.

Zatorre, R. J., & Salimpoor, V. N. (2013). From perception to pleasure: Music and its neural substrates. *Proceedings of the National Academy of Sciences of the United States of America, 110*, S10430–10437.

Zendel, B. R. & Alain, C. (2014). Enhanced attention-dependent activity in the auditory cortex of older musicians. *Neurobiology of Aging, 35,* 55–63.

Ziv, N., Granot, A., Hai, S., Dassa, A., & Haimov, I. (2007). The effect of background stimulative music on behavior in Alzheimer's patients. *Journal of Music Therapy, 4,* 329–343.

# 7

# Musical Playlists for Addressing Depression in People With Dementia

*Sandra Garrido*

## The Relationship Between Depression and Dementia

Many people with dementia also commonly experience depression. In fact, dementia and depression can be difficult to distinguish because of the overlap of symptoms. Stories of older adults who become strangely withdrawn and depressed, and who are finally diagnosed as being in the early stages of dementia, are heard all too frequently. As many as 20% of people with Alzheimer's disease (AD) and 30% of people with dementia with Lewy bodies (DLB) may also have Major Depressive Disorder (MDD; Whitfield et al., 2015). Similarly, 33% of people with frontotemporal dementia (FTD) have depressive symptoms (Chakrabarty, Sepehry, Jacova, & Hsiung, 2015), and 63% of people with Mild Cognitive Impairment (MCI) may also have MDD (Panza et al., 2010).

Increasing loss of independence, mobility, and the biological changes that occur in people with dementia can contribute to this accompanying depression. The slow corrosion of cognitive functioning, reduced capacity to perform the tasks necessary to everyday life, and diminishing social contact can wear away an individual's sense of self and self-confidence. In addition to the lowered mood that can occur as a result of these changes, a more complex link between depression and dementia exists. Along with the high rates of co-morbidity between depression and dementia, there is also strong evidence of a bi-directional relationship, with studies suggesting that depression is both a risk factor and a prodromal symptom of dementia (Azner & Knudsen, 2011). Depressive episodes in later life can increase the risk of developing dementia by up to 3 times (Brommelhoff et al., 2009), and people with MCI are twice as likely to develop AD if they also have MDD (Ownby, Crocco, Acevedo, John, & Loewensteing, 2006). In addition, the presence of depression seems to increase rates of cognitive and functional decline as well as hasten institutionalization in people with dementia (Potter & Steffens, 2007). Thus it has been proposed that depression may demonstrate a susceptibility to dementia, as well as being an early symptom of dementia (Kasahara et al., 2006).

Given the frequent co-occurrence of depression and dementia, some argue that the neuropsychological changes that occur with recurrent depressive episodes increase the brain's vulnerability to the degenerative processes involved in dementia (Aznar & Knudsen, 2011). For example, depression is well known to involve dysfunctions in serotonin neurotransmission, with AD similarly being associated with reduced levels of serotonin and serotonergic neurons (Premi et al., 2015). Some studies suggest that vascular disease may be the primary linking factor between depression and dementia (Alexopoulos, 2003). Alternatively, it is theorized that high cortisone levels—which are frequently observed in depression and are also associated with hippocampal atrophy in AD—provide the connection (van de Pol et al., 2006).

Given these numerous overlaps in psychopathology it is no surprise that both dementia and MDD are increasingly prevalent worldwide. MDD is currently ranked by the World Health Organization as the number one contributor to nonfatal health loss, and it is predicted that both MDD and dementia will be among the diseases causing the greatest societal burdens during the 21st century. The close connection between the two diseases has led some to argue that a 25% reduction in depression prevalence could result in 827,000 fewer cases of AD worldwide, the most prevalent form of dementia (Bames & Yaffe, 2011). In any case, further investigation of the biological mechanisms underpinning both conditions is needed in order to fully understand their etiology as well as appropriate treatment (Leyhe et al., 2017).

According to both people with dementia and their caregivers, depression has a substantial impact on their quality of life (Banerjee et al., 2009), compounding the stress upon both caregivers and health services (Curran & Loi, 2012). This is complicated by the fact that depression in people with dementia can be difficult to diagnose since many of the symptoms associated with depression such as difficulties with concentration and reduced interest in normally enjoyed activities can also be symptoms of cognitive decline (Pattanayak & Sagar, 2011).

Even when accurately diagnosed, depression in people with dementia is often untreated or mismanaged (Thyrian et al., 2016), with only about one third of sufferers receiving effective treatment (Gottfries, 2001). The presence of multiple co-morbidities and high levels of medication being prescribed for other conditions makes pharmacological approaches problematic in this population (Gottfries, 2001). Where antidepressants are used for people with dementia, for example, the evidence suggests that they are not always effective in diminishing symptoms of depression and can at times be associated with adverse clinical events such as increased confusion (Farina, Morrell, & Banerjee, 2017).

## Music and Depression in the General Population

### Improvement of Mood

Mood regulation is one of the primary reasons people listen to music (Garrido & Schubert, 2011; Saarikallio, 2008). Music listening is also one of the most powerful means for improving mood (Thayer, Newman, & McClain, 1994). Music is thus often used for addressing depression in populations without dementia, both young and old. As outlined in Chapter 2 of this volume, music therapy includes both active engagement with music using interventions such as song-writing, improvisation, and group singing, and receptive forms such as listening to live or pre-recorded music (Guetin et al., 2009). Music can also be used for therapeutic purposes outside of formal music therapy, both in healthcare contexts and by individuals in their everyday lives.

In younger individuals, music listening has been found to increase well-being in mothers with postnatal depression (Fancourt & Perkins, 2017), to reduce symptoms in female students with depression (Esfandiari & Mansouri, 2014), and to alleviate depression in people with neurological disorders such as stroke, epilepsy, multiple sclerosis and Parkinson's disease (Raglio, Attardo, et al., 2015). Music listening can benefit mental health by providing opportunities for emotion modification and emotional immersion (Papinczak, Dingle, Stoyanov, Hides, & Zelenko, 2015). Studies have also shown that listening to music can reduce depression in older adults living in the community compared to a control group (Chan, Chan, Mok, & Tse, 2009; Chan, Wong, Onishi, & Thayala, 2012). Music-therapy sessions conducted by a trained music therapist have also been found to reduce depressive symptoms among older adults (Zhao, Bai, Bo, & Chi, 2016).

## Worsening of Mood

However, the relationship between music use and depression is complex. Research has shown that people with depression are often attracted to music that can worsen their symptoms (Garrido & Schubert, 2013). Numerous studies have shown that young people in particular may tend to listen to music that does not always support well-being, and may even report feeling worse after listening to music (Garrido & Schubert, 2015; Miranda & Claes, 2009; Skewes McFerran & Saarikallio, 2014; Wilhelm, Gilllis, Schubert, & Whittle, 2013).

The negative effects that music can have on people with depression stem from the very nature of the disorder. Depression is by definition a disorder of affect dysregulation, meaning that people with depression lack effective strategies for successfully managing their moods, often engaging in behavior that can prolong negative affective states (Forbes & Dahl, 2005). People with depression tend to have a bias towards negative stimuli, frequently interpreting it as more negative than people without depression. They are also inclined to give more attention to and to remember negative stimuli more than positive stimuli. For example, studies have shown that people with depression are both more likely to interpret facial expressions as conveying negative emotions (Raes, Hermans, & Williams, 2006) and to show greater attention to sad faces than to happy or angry faces (Gotlib, Krasnoperova, Neubauer Yue, & Joormann, 2004).

Neuroimaging studies using functional magnetic resonance imaging (fMRI) have shown that people with depression show sustained blood flow in the amygdala—the brain area responsible for emotion coding (Siegle, Steinhauer, Thase, Stenger, & Carter, 2002). Participants with depression displayed ongoing activity in that brain region when processing negatively valenced stimuli, in contrast to never-depressed participants whose amygdala responses decayed after stimuli presentation. Thus people with depression appear to have a reduced ability to modulate amygdala activity through cognitive processes, suggesting a neurological basis for their negative attentional bias. This attentional bias applies to musical stimuli too, with several studies reporting that people with depression showed a heightened response to sad musical excerpts, selecting more descriptive labels in response to sad music than other excerpts in comparison to a control group (Bodner et al., 2007), or a tendency to evaluate music more negatively than healthy controls (Punkanen, Eerola, & Erkkila, 2011).

In addition to being more sensitive to stimuli evoking negative emotions, people with depression also tend to have impaired mechanisms for mood regulation and to lack the cognitive control needed to successfully disengage from negative information (Koster, Lissnyder, Derakshan, & Raedt, 2011; Vanderhasselt et al., 2012; Vanderhasselt, Kuhn, & Raedt, 2011). As discussed above, this appears to occur at a neurological level, with amygdala responses to negative musical stimuli being heightened and sustained in people with depression, in contrast to nondepressed individuals for whom amygdala responses decay relatively quickly (Siegle et al., 2002). This means that when negative affective states are induced, by music or otherwise, people with depression have more trouble recovering from these states.

Much of this may have to do with the ingrained thinking patterns and coping styles of the individual. Rumination, for example, is a largely involuntary compulsion to focus on negative and pessimistic thoughts which is highly predictive of clinical depression (Joorman, 2005). Rumination is distinct from self-reflection, since the latter allows an individual to process and work through negative emotions resulting in an improved mood, while the former involves cyclical patterns of negative thinking from which people with depression can have difficulty emerging.

Since people with tendencies to depression have a propensity to perceive sad music as even more negative than other people and also find it difficult to disengage from the negative emotions aroused by it, they often do not benefit from listening to music in the same way as other individuals. Studies have shown that music can exacerbate the cycles of negative thinking in which ruminators frequently become stuck, tending to induce negative memories and thought processes more than in non-ruminators (Garrido, Bangert, & Schubert, 2016), and at the same time deepening their depressed mood (Garrido & Schubert, 2015). A person with no history of depression, therefore, might listen to a piece of sad music and not experience sadness, recover quite quickly from the emotions induced, or use the opportunity to process and work through negative emotions. In contrast, a person with tendencies to depression, or those who are currently experiencing depression, may be more likely to experience negative affective states in response to music and find it more difficult to recover from the negative affective state induced. Thus, people with symptoms of or a history of depression may need special care taken when music interventions are used. The incidence of similar issues with regards to music use in people with both depression and dementia has not been explored to the same degree in the literature.

## Music and Depression in People With Dementia

Numerous studies have demonstrated that active forms of music therapy can have a positive influence on depressive symptoms, agitation, and cognitive functioning in people with dementia (O'Connor, Ames, Gardner, & King, 2009). Despite these obvious benefits, musical interventions for people with dementia that require the presence of a professional therapist or live musician are limited by significant costs and logistical factors, making them inaccessible to many patients (Nair, Browne, Marley, & Heim, 2013). There is thus increasing interest in the use of prerecorded music for therapeutic purposes outside of formal music-therapy settings.

Such interventions have the advantage of being relatively easy to access and affordable. Prerecorded music is widely available to most people whether in residential or home-care situations, and it can be accessed at the time it is most needed, as frequently as needed. However, these types of interventions do not involve a qualified musical therapist and therefore lack the advantage of the patient-therapist relationship which is crucial to the practice of music therapy (Raglio, Bellandi, et al., 2015).

Studies of music listening in people with dementia outside formal music therapy, therefore, tend to return mixed results. One reason for this may be the fact that the music played is often researcher-selected. For example, under the assumption that instrumental or classical music is relaxing, some studies have examined the effects of playing a single type of music such as Baroque music or nature music to groups of older people with dementia and have found that participants may even demonstrate an increase in agitated behavior (see for example Chang, Huang, Lin, & Lin, 2010). It is understandable that an individual who dislikes Baroque music, for example, would experience a deterioration of mood after being exposed to it for a long period. Thus, there is increasing recognition that self-selected music and music that is biographically salient is more effective as a mood moderator than music that an individual has no particular affinity with and may even actively dislike (Nair et al., 2013).

## Improvement of Mood

Given the increasingly obvious fact that people with dementia will be more likely to respond positively to music that is personally relevant to them or of a style or genre that they prefer, interest has begun to shift towards the use of personalized playlists in healthcare contexts. "Personalized playlists" refers to the creation of playlists of prerecorded music by researchers, health-care workers, or caregivers based on the tastes of the individual listener. Information about the preferences of the individual and about specific songs that may hold particular autobiographical significance for the individual is generally based on details provided by the individual themselves or friends and family. An established protocol for gathering this information may be used, such as that developed by Linda Gerdner (Gerdner, 2000; Gerdner, Hartsock, & Buckwalter, 2000), or music choices may just be based on ge-neral information about the individual's "favorite music." The selected music is then played to the individual for therapeutic purposes.

Despite the obvious advantages of personalizing the music played to an individual with dementia, studies attempting to confirm the efficacy of these interventions also show mixed results, particularly in addressing depression. In a recent review of 28 studies using prerecorded music with people with dementia, my colleagues and I (Garrido et al., 2017) found only one study that demonstrated a positive impact on depression from individualized music listening interventions that did not directly involve a music therapist. A more recent evaluation of one personalized playlist intervention also re-ported a lack of significant improvements in depression symptoms (Kwak, Brondino, O'Connell Valuch, & Maeda, 2016). Other studies, while reporting that a majority of participants benefited from the intervention, acknowl-edged that these results were not universal, with some responding neutrally or even experiencing a worsening of mood (Martin, Schroeder, Smith, & Jones, 2016).

What could account for these diverse responses even when music is individualized to the recipient's own taste? Firstly, as noted by Kwak and colleagues (2016), music listening programs are most likely to be beneficial to individuals for whom music has had some importance in their lives. Such interventions are unlikely to have positive effects on individuals who are disinterested in or show a dislike for listening to music. Further important considerations are the symptoms of the individual and their mental health history. While an increased understanding of the role of individual music preferences is shifting interventions in a positive direction towards greater individualization, one result of this is a decreasing consideration of the effect of particular music on individual symptoms. Few interventions considered in the literature appear to be able to integrate a personalized approach while also targeting the music towards the symptoms for which relief is sought. This is particularly important where the symptoms sought to be addressed are related to mood and emotion.

For example, people who are agitated might in some circumstances benefit from music that is soothing and relaxing. In contrast, people who are depressed might tend to find the same music sad and gloomy, with a correspondingly negative impact on their mood after listening. Thus, as noted by Kwak and colleagues (2016), it is vital to identify the most prominent and distressing symptoms, and develop the playlist accordingly. Kwak and colleagues thus argue that playlists for people whose primary symptoms are agitation or aggression may need to "be developed with the goal to help the resident relax and increase positive mood," while for those who are unmotivated or withdrawn the goal may need to be "to help the resident be energized and actively engaged with the environment" (p. 71).

One notable study which did attempt to consider both individual preferences and the desired mood effects involved a music-therapy intervention for people with dementia using specially created instrumental music in several styles (Guetin et al., 2009). While participants could select their preferred genre, all the musical sequences followed a predetermined "U-shaped" curve in each listening session, first lowering arousal by reducing the tempo, orchestral texture, and volume and then increasing it. Significant reductions in anxiety and depression were found. Since this study included only participants with high anxiety, it is understandable that music designed to lower arousal would have a relaxing effect on participants. It is important to note, however, that both dementia and depression can present with varying levels of arousal including high arousal states like anxiety, or low arousal states such as apathy or catatonia (Starkstein & Pahissa, 2014). It is unclear whether music designed to lower arousal would have a similar beneficial effect on patients with differing symptomatic profiles. Therefore, whether the arousal potential of the music and the arousal needs of the participant are well matched is a further factor needing consideration.

Where an individual is evincing clinical levels of depression or has a history of it, even more care may need to be taken when implementing music interventions. As discussed in detail above, people with depression are both more likely to interpret a piece of music as negative and to have difficulty in recovering from a negative mood once it is induced by listening to a piece of music. Individuals with a history of depression or who are currently experiencing it may thus be at risk for adverse reactions to music, with the possibility that their mood might actually become worse after listening to it. Care must therefore be taken to ensure that music selected for such at-risk individuals is unlikely to stimulate negative memories or trigger negative thought patterns to which such individuals are prone.

Merely playing the "favorite music" of the individual may not be enough to guarantee that the desired mood improvement is achieved. While sad or nostalgic music may confer psychological benefits on healthy participants, the same music can exacerbate patterns of negative thinking, resulting in mood deterioration in people with depression (Garrido, 2018; Garrido & Schubert, 2015). Depressed people also demonstrate a low level of awareness of the impact their self-selected music has on their mood, being habitually attracted to music that perpetuates dysphoria. Thus people with depression may actually have a long-term history of listening to music that worsens their mood and of using music within maladaptive patterns of coping behaviors (Garrido & Schubert, 2013). The flawed assumption that music will affect all listeners in the same way may underpin many of the anomalies in previous studies.

## Worsening of Mood

While it certainly seems true that people *without* dementia may experience mood deterioration after listening to music, it may in fact be the case that people *with* dementia are even more vulnerable to adverse responses. In general, research has demonstrated that the tendency towards a negative attentional bias seems to decrease with age (Kennedy, Mather, & Cartensen, 2004). Older adults show a greater preference for positive information in both attention and memory than younger people (Mikels, Larkin, Reuter-Lorenz, & Cartensen, 2005). Numerous studies have demonstrated a similar "positivity effect" in people with dementia, in that people with AD were found to recall a greater number of positive music-evoked memories than younger counterparts (Cuddy, Sikka, Silveira, Bai, & Vanstone, 2017) Nevertheless, it takes some degree of cognitive control to direct attention and processing towards positive stimuli (Reed & Carstensen, 2012). Thus, as cognitive decline increases with the advance of dementia, the ability to redirect thoughts away from negative material can also decline. In fact, one study found that people with AD remembered a greater proportion of negative versus positive words compared to either healthy younger or older adults (Fleming, Kim, Doo, Maguire, & Potkin, 2003), suggesting that the capacity for cognitive moderation of thoughts and emotions declines in tandem with general cognitive ability in people with dementia. Thus, people with dementia who have a history of depression may prove to be even more susceptible to the effects of music triggering negative thoughts or negative memories than depressed individuals without dementia.

The vulnerability of individuals with dementia and particularly those with depression was borne out by some recent studies conducted by myself and my colleagues (Garrido, Stevens, Chang, Dunne, & Perz, 2018). The research involved compiling individual playlists for 99 people with various types of dementia. Music was selected on the basis of participants' personal preferences and categorized according to tempo (slow and fast), mode (major or minor), and lyrics (no lyrics, negative content, and positive content) in order to determine whether particular musical features interacted in specific ways with the symptoms of the listener. One of our key findings was that people with high levels of depression as measured by the Patient Health Questionnaire (Spitzer, Kroenke, & Williams, 1999) demonstrated increased activation of facial muscles associated with sadness after listening to music regardless of the features of music they listened to (Garrido, Stevens, et al., 2018). On the other hand we found that people with dementia who had low levels of depression but were apathetic and withdrawn were the most likely to demonstrate behavioural evidence of pleasure during music listening such as foot tapping, humming, smiling, or singing along. This suggests that while people with clinical levels of depression or a history of it were particularly vulnerable to negative affective responses, those perhaps at preclinical levels of depression with symptoms of apathy and social detachment, were among those who benefited most from music listening. We also found that people with symptoms of Alzheimer's disease were more likely to experience these negative affective responses than people with symptoms of dementia with Lewy bodies. Our research further demonstrated that music in minor modes may be particularly likely to increase negative affective responses, while music in fast tempos tends to increase arousal levels as suggested by galvanic skin response (Garrido, Stevens, Chang, Dunne, & Perz, 2019). Self-reported mood ratings indicated that this increased arousal was unpleasant for participants, something akin to overstimulation.

Despite the negative response to music from participants with depression in our study, it need not be assumed that music interventions cannot be of any benefit to such individuals. Rather, these findings highlight the need to identify those particularly at risk of adverse responses prior to commencing any type of music intervention. Music for such vulnerable individuals, particularly those with clinical levels or a history of depression, may need to be selected with higher levels of care. As in research with younger participants with depression, it is likely that music that connects people with positive memories or that they particularly associate with positive thoughts and happy times, may serve to improve mood in these vulnerable individuals even more than in the average listener (Garrido & Schubert, 2015). Higher levels of monitoring may also be necessary during the implementation of music interventions in people with dementia with clinical depression, in order to determine what works best for the individual. Where possible, more formal music therapy with a trained therapist may also be useful.

At present, as observed by Kwak et al. (2016), even facilities following trademarked programs such as the Music and Memory program as popularized in the film *Alive Inside* (Rossato-Bennett, 2014) tend to fail to target the music towards the symptoms of the individual or to provide appropriate monitoring. Music selections may be personalized, but the time, place, and mode of delivery may often follow standard routines within a care facility, failing to account for individual symptoms or the diurnal variations of those symptoms. Furthermore, recent research suggests that in many residential aged care facilities, communal consumption of prerecorded music considered "age-appropriate" is still one of the most common forms of musical engagement (Garrido, Dunne, Perz, Chang, & Stevens, 2018). This highlights the need for standardized guidelines for music selection for caregivers and aged care workers that account for individual variables beyond merely musical taste. Further research will be required to increase an understanding of how mental health history contributes to individual responses to music in people with dementia. In addition, much of the research that has been conducted about dementia and music has been conducted in people with Alzheimer's disease. Given important differences in the symptoms and pathology between different types of dementia, there is a great need for further research to investigate the influence that the type of dementia has upon response to musical interventions.

# References

Alexopoulos, G. S. (2003). Vascular disease, depression, and dementia. *Journal of the American Geriatric Society, 51*, 1178–1180.

Azner, S., & Knudsen, G. M. (2011). Depression and Alzheimer's disease: Is stress the initiating factor in a common neuropathological cascade? *Journal of Alzheimer's Disease, 23*, 177–193.

Bames, D. E., & Yaffe, K. (2011). The projected effect of risk factor reduction on Alzheimer's disease prevalence. *Lancet Neurology, 10*, 819–828.

Banerjee, S., Samsi, K., Petrie, C. D., Alvir, J., Treglia, M., Schwam, E. M., & del Valle, M. (2009). What do we know about quality of life in dementia? A review of the emerging evidence on the predictive and explanatory value of disease specific measures of health related quality of life in people with dementia. *International Journal of Geriatric Psychiatry, 24*(1), 15–24.

Bodner, E., Iancu, I., Gilboa, A., Sarel, A., Mazor, A., & Amir, D. (2007). Finding words for emotions: The reactions of patients with major depressive disorder towards various musical excerpts. *The Arts in Psychotherapy, 34*, 142–150.

Brommelhoff, J., Gatz, M., Johansson, B., McArdle, J., L., F., & Pedersen, N. (2009). Depression as a risk factor or prodromal feature for dementia? Findings in a population-based sample of Swedish twins. *Psychology and Aging, 24*, 373–384.

Chakrabarty, T., Sepehry, A. A., Jacova, C., & Hsiung, G.-Y. R. (2015). The prevalence of depressive symptoms in frontotemporal dementia: A meta-analysis. *Dementia and Geriatric Cognitive Disorders, 39*, 257–271.

Chan, M. F., Chan, E. A., Mok, E., & Tse, F. Y. K. (2009). Effect of music on depression levels and physiological responses in community-based older adults. *International Journal of Mental Health Nursing, 18*(4), 285–294.

Chan, M. F., Wong, Z. Y., Onishi, H., & Thayala, N. V. (2012). Effects of music on depression in older people: A randomised controlled trial. *Journal of Clinical Nursing, 21*(5-6), 776–783.

Chang, F. Y., Huang, H. C., Lin, K. C., & Lin, L. C. (2010). The effect of a music programme during lunchtime on the problem behaviour of the older residents with dementia at an institution in Taiwan. *Journal Clinical Nursing, 19*(7–8), 939–948. doi:10.1111/j.1365-2702.2009.02801.x

Cuddy, L., Sikka, R., Silveira, K., Bai, S., & Vanstone, A. (2017). Music-evoked autobiographical memories (MEAMs) in Alzheimer's disease: Evidence for a positivity effect. *Cogent Psychology, 4*(1).

Curran, E. M., & Loi, S. (2012). Depression and dementia. *Medical Journal of Australia, 1*(Suppl 4), 40–44.

Esfandiari, N., & Mansouri, S. (2014). The effect of listening to light and heavy music on reducing symptoms of depression among female students. *The Arts in Psychotherapy, 41*(2), 211–213.

Fancourt, D., & Perkins, R. (2017). Associations between singing to babies and symptoms of postnatal depression, well-being, self-esteem, and mother-infant bond. *Public Health, 145*, 149–152.

Farina, N., Morrell, L., & Banerjee, S. (2017). What is the therapeutic value of antidepressants in dementia? A narrative review. *International Journal of Geriatric Psychiatry, 32*(1), 32–49.

Fleming, K., Kim, S. H., Doo, H., Maguire, G., & Potkin, S. G. (2003). Memory for emotional stimuli in patients with Alzheimer's disease. *American Journal of Alzheimer's Disease and Other Dementias, 18*, 340–342. doi:10.1037/0882-7974.5.3.335

Forbes, E. E., & Dahl, R. C. (2005). Neural systems of positive affect: Relevance to understanding child and adolescent depression? *Development and Psychopathology, 17,* 827–850.

Garrido, S. (2018). The influence of personality and coping style on the affective outcomes of nostalgia: Is nostalgia a healthy coping mechanism or rumination? *Personality and Individual Differences, 120,* 259–264. doi:10.1016/j.paid.2016.07.021

Garrido, S., Bangert, D., & Schubert, E. (2016). Musical prescriptions for mood improvements: A mixed methods study. *The Arts in Psychotherapy, 51,* 46–53.

Garrido, S., Dunne, L., Chang, E., Perz, J., Stevens, C., & Haertsch, M. (2017). The use of music playlists for people with dementia: A critical synthesis. *Journal of Alzheimer's Disease, 60,* 1129–1142. doi:10.3233/JAD-170612

Garrido, S., Dunne, L., Perz, J., Chang, E., & Stevens, C. (2018). The use of music in aged care facilities: A mixed methods study. *Journal of Health Psychology, 15,* 765–776.

Garrido, S., & Schubert, E. (2011). Negative emotion in music: What is the attraction? A qualitative study. *Empirical Musicology Review, 6*(4), 214–230.

Garrido, S., & Schubert, E. (2013). Adaptive and maladaptive attraction to negative emotion in music. *Musicae Scientiae, 17*(2), 145–164. doi:10.1177/1029864913478305

Garrido, S., & Schubert, E. (2015). Music and people with tendencies to depression. *Music Perception, 32*(4), 313–321. doi:10.1525/MP.2015.32.4.313

Garrido, S., Stevens, C., Chang, E., Dunne, L., & Perz, J. (2018). Music and dementia: Individual differences in response to personalized playlists. *Journal of Alzheimer's Disease, 64*(3), 933–941. doi:10.3233/JAD-180084

Garrido, S., Stevens, C., Chang, E., Dunne, L., & Perz, J. (2019). Music and dementia: Musical features and affective responses to personalized playlists. *American Journal of Alzheimers and Other Dementias, 34*(4), 247–253.

Gerdner, L. A. (2000). *Evidence-based protocol: Individualized music intervention.* Retrieved from Iowa City: IA.

Gerdner, L. A., Hartsock, J., & Buckwalter, K. C. (2000). *Assessment of Personal Music Preference (Patient Version).* University of Iowa. Iowa. Retrieved from http://www.carepartnermentoring.com/music/Music Preference Questionnaire Patient Version.pdf

Gotlib, I. H., Krasnoperova, E., Neubauer Yue, D., & Joormann, J. (2004). Attentional biases for negative interpersonal stimuli in clinical depression. *Journal of Abnormal Psychology, 113*(1), 127–135.

Gottfries, C.-G. (2001). Late life depression. *European Archives of Psychiatry and Clinical Neuroscience, 251*(Suppl 2), 57–61.

Guetin, S., Porter, F., Picot, M. C., Pommie, C., Messaoudi, M., Djabelkir, L., . . . Touchon, J. (2009). Effect of music therapy on anxiety and depression in patients with Alzheimer's type dementia: randomised, controlled study. *Dement Geriatr Cogn, 28*(1), 36–46.

Joorman, J. (2005). Inhibition, rumination and mood regulation in depression. In R. W. Engle, G. Sedek, U. v. Hecker, & D. N. McIntosh (Eds.), *Cognitive limitations in aging and psychopathology: Attention, working memory, and executive functions* (pp. 275–312). New York: Cambridge University Press.

Kasahara, H., Tsumura, M., Ochiai, Y., Furukawa, H., Aoki, K., Ito, T. A., Kada, H., Hasimude, T., & Nakanish, T. (2006). Consideration of the relationship between depression and dementia. *Psychogeriatrics, 6,* 128–133.

Kennedy, Q., Mather, M., & Cartensen, L. L. (2004). The role of motivation in the age-related positivity effect in autobiographical memory. *Psychological Science, 15*(208–214). doi:10.1111/j.0956-7976.2004.01503011.x

Koster, E. H. W., Lissnyder, E. D., Derakshan, N., & Raedt, R. D. (2011). Understanding depressive rumination from a cognitive science perspective: The impaired disengagement hypothesis. *Clinical Psychology Review, 31*(1), 138–145.

Kwak, J., Brondino, M. J., O'Connell Valuch, K., & Maeda, H. (2016). *Evaluation of the Music and Memory program among nursing home residents with dementia: Final report to the Wisconsin Department of Health Services.* Retrieved from Wisconsin: <https://www.dhs.wisconsin.gov/publications/p01594.pdf>

Leyhe, T., Reynolds, C. F., Melcher, T., Linnemann, C., Kloppel, S., Blennow, K., ... Hampel, H. (2017). A common challenge in older adults: Classification, overlap, and therapy of depression and dementia. *Alzheimer's & Dementia, 13*, 59–71.

Martin, P. K., Schroeder, R. W., Smith, J. M., & Jones, B. (2016). The Roth project—Music and Memory: Surveying the observed benefit of personalized music in individuals with diagnosed or suspected dementia. *Alzheimer's & Dementia, 12*(7 Supplement), P988. doi:http://dx.doi.org/10.1016/j.jalz.2016.06.2028

Mikels, J. A., Larkin, G. R., Reuter-Lorenz, P. A., & Cartensen, L. L. (2005). Divergent trajectories in the aging mind: Changes in working memory for affective versus visual information with age. *Psychology and Aging, 20*, 524–553. doi:10.1037/0882-7974.20.4.542

Miranda, D., & Claes, M. (2009). Music listening, coping, peer affiliation and depression in adolescence. *Psychology of Music, 37*(2), 215–233.

Nair, B. R., Browne, W., Marley, J., & Heim, C. (2013). Music and dementia. *Degenerative Neurological and Neuromuscular Disease, 3*, 47–51.

O'Connor, D. W., Ames, D., Gardner, B., & King, M. (2009). Psychosocial treatments of behavior symptoms in dementia: A systematic review of reports meeting quality standards. *International Psychogeriatrics, 21*(2), 225–240.

Ownby, R., Crocco, E., Acevedo, A., John, V., & Loewenstein, D. (2006). Depression and risk for Alzheimer disease: Systematic review. meta-analysis, and metaregression analysis. *Archives of General Psychiatry, 63*, 530–538.

Panza, F., Frisardi, V., Capurso, C., D'Introno, A., Colacicco, A., Imbimbo, B., ... Solfrizzi, V. (2010). Late-life depression, mild cognitive impairment, and dementia: Possible continuum? *American Journal of Geriatric Psychiatry, 18*, 98–116.

Papinczak, Z. E., Dingle, G. A., Stoyanov, S. R., Hides, L., & Zelenko, O. (2015). Young people's uses of music for well-being. *Journal of Youth Studies, 18*(9), 1119–1134.

Pattanayak, R., & Sagar, R. (2011). Depression in dementia patients: Issues and challenges for a physician. *Journal of Associated Physicians India, 59*, 650–652.

Potter, G., & Steffens, D. (2007). Contribution of depression to cognitive impairment and dementia in older adults. *Neurologist, 13*, 105–117.

Premi, E., Silvana, A., Pilotto, A., Seripa, D., Paghera, B., Padovani, A., & Borroni, B. (2015). Functional genetic variation in the serotonin 5-HTTLPR modulates brain damage in frontotemporal dementia. *Neurobiology of Aging, 36*(1), 446–451.

Punkanen, M., Eerola, T., & Erkkila, J. (2011). Biased emotional preferences in depression: Decreased liking of angry and energetic music by depressed patients. *Music and Medicine, 3*(2), 114–120.

Raes, F., Hermans, D., & Williams, J. M. G. (2006). Negative bias in the perception of others' facial emotional expressions in major depression: The role of depressive rumination. *Journal of Nervous and Mental Disease, 194*(10), 796–799.

Raglio, A., Attardo, L., Gontero, G., Rollino, S., Groppo, E., & Granieri, E. (2015). Effects of music and music therapy on mood in neurological patients. *World Journal of Psychiatry, 5*(1), 68–78.

Raglio, A., Bellandi, D., Baiardi, P., Gianotti, M., Ubezio, M. C., Zanacchi, E., . . . Stramba-Badiale, M. (2015). Effect of active music therapy and individualized listening to music on dementia: A multicenter randomized controlled trial. *Journal of the American Geriatrics Society, 63*(8), 1534–1539. doi:10.1111/jgs.13558

Reed, A. E., & Carstensen, L. L. (2012). The theory behind the age-related positivity effect. *Frontiers in Psychology, 3*(339). doi:http://doi.org/10.3389/fpsyg.2012.00339

Rossato-Bennett, M. (Writer). (2014). *Alive Inside: A story of music and memory*. USA.

Saarikallio, S. (2008). Music in mood regulation: Initial scale development. *Musicae Scientiae, 12*(2), 291–309.

Siegle, G. J., Steinhauer, S. R., Thase, M. E., Stenger, A., & Carter, C. S. (2002). Can't shake that feeling: Event-related fMRI assessment of sustained amygdala activity in response to emotional information in depressed individuals. *Biological Psychiatry, 51*, 693–707.

Skewes McFerran, K., & Saarikallio, S. (2014). Depending on music to feel better: Being conscious of responsibility when appropriating the power of music. *The Arts in Psychotherapy, 41*(1), 89–97.

Spitzer, R. L., Kroenke, K., & Williams, J. B. (1999). Patient Health Questionnaire Study Group. Validity and utility of a self-report version of PRIME-MD: The PHQ Primary Care Study. *JAMA, 282*, 1737–1744.

Starkstein, & Pahissa. (2014). Psychiatric complications of Alzheimer's disease overlapping with Parkinsonism: Depression, apathy, catatonia, and psychosis. In M. Marello & S. Starkstein (Eds.), *Movement Disorders in Dementias* (pp. 73–86). London: Springer-Verlag.

Thayer, R. E., Newman, R., & McClain, T. M. (1994). Self-regulation of mood: Strategies for changing a bad mood, raising energy, and reducing tension. *Journal of Personality and Social Psychology, 67*, 910–925.

Thyrian, J., Eichler, T., Reimann, M., Wucherer, D., Dreier, A., Michalowsky, B., & Hoffmann, W. (2016). Depressive symptoms and depression in people screened positive for dementia in primary care—results of the DelpHi-study. *International Psychogeriatrics, 28*(6), 929–937. doi:10.1017/S1041610215002458

van de Pol, L., Hensel, A., Barkhof, F., Gertz, H., Scheltens, P., & van der Flier, W. (2006). Hippocampal atrophy in Alzheimer disease: Age matters. *Neurology, 66*, 236–238.

Vanderhasselt, M.-A., Kuhn, S., & Raedt, R. D. (2011). Healthy brooders employ more attentional resources when disengaging from the negative: An event-related fMRI study. *Cognitive, Affective & Behavioral Neuroscience, 11*(2), 207–216.

Vanderhasselt, M.-A., Raedt, R. D., Dillon, D. G., Dutra, S. J., Brooks, N., & Pizzagalli, D. A. (2012). Decreased cognitive control in response to negative information in patients with remitted depression: An event-related potential study. *Journal of Psychiatry and Neuroscience, 37*(4), 250–258.

Whitfield, D. R., Vallortigara, J., Alghamdi, A., Hortobagyi, T., Ballard, C., Thomas, A. J., . . . Francis, P. T. (2015). Depression and synaptic zinc regulation in Alzheimer disease, dementia with Lewy bodies, and Parkinson disease dementia. *The American Journal of Geriatric Psychiatry, 23*(2), 141–148.

Wilhelm, K., Gilllis, I., Schubert, E., & Whittle, E. L. (2013). On a blue note: Depressed people's reasons for listening to music. *Music and Medicine, 5*(2), 76–83.

Zhao, K., Bai, Z. G., Bo, A., & Chi, I. (2016). A systematic review and meta-analysis of music therapy for the older adults with depression. *International Journal of Geriatric Psychiatry, 31*(11), 1188–1198.

# 8

# Preserved Musical Instrument Playing in Dementia

## A Unique Form of Access to Memory and the Self

*Amee Baird and William Forde Thompson*

The ability to play a musical instrument can remain despite significant cognitive impairment in the severe stage of dementia. Such observations illustrate a striking dissociation between musical and non-musical memory and skill within this population. What are the psychosocial implications of such a complex skill being preserved in the face of an otherwise devastating cognitive impairment? We propose that playing a musical instrument allows a unique form of access to two crucial domains: memory and the self. This notion was first raised by Matthews (2015), who described music as a "tool of access—access to memory of music and access to a past social self" in people with dementia (p. 576). We expand this notion by considering musical instrument playing as a means of accessing specific types of memory and maintaining a strong sense of self. In particular, certain procedural, semantic, and episodic, including autobiographical, memories may be preserved through the act of playing an instrument, even in the severe stage of dementia. Such memories are not easily elicited by non-music stimuli (e.g., verbal and visual reminders), but playing a musical instrument has a unique capacity to enable the expression of these preserved forms of memory. In doing so, it can provide access to one's past and continuing self, and can be considered a form of self-preservation and expression in musicians with dementia.

In this chapter, we review existing literature on preserved musical instrument playing in people with dementia, which comprises single case studies or small case series of people with Alzheimer's Dementia (AD) or different variants of frontotemporal dementia (FTD). We also describe four new cases of preserved musical instrument playing in people with AD and Bv-FTD, and discuss evidence that musical instrument playing enables unique access to different types of memory and the self.

## Playing a Music Instrument

Playing a musical instrument is a highly complex skill that involves the integration of numerous cognitive, psychomotor, and affective functions. It is a multimodal activity that involves an integrated network of aural, visual, tactile and motor skills, along with detailed procedural memory and emotional sensitivity. Music performance requires retrieval of musical structures and units of knowledge from memory (e.g., melodies, harmonic progressions, rhythmic groupings, metric templates), followed by motor planning, preparation, and production of appropriately timed movements (see Palmer, 1997 for a review). Fine motor and bimanual coordination, speed, timing, motor sequencing, and intact visual perception (for reading musical notation) and auditory perception (for auditory motor feedback) are all crucial skills for playing a musical instrument, regardless of the type. Musical performance activates widespread neural networks, including primary sensory and motor regions, association areas, frontal brain regions, and emotional and memory networks. As suggested by Fornazzari et al. (2017), practicing a musical instrument over time may facilitate more efficient interactions between these brain regions, which enables compensation for deficits associated with neurodegeneration. In this way, music engagement may have protective benefits for dementia, and provides a scaffold for the range of skills that it engages.

There is extensive evidence that music training elicits widespread brain plasticity, especially in sensory and motor brain regions (Herholz & Zatorre, 2012; Omigie & Samson, 2014). Much of this research has been conducted on young musicians, but several studies have documented functional brain changes even after a short period of music training in older adults (Bugos, Perlstein, McCrae, Brophy, & Bedenbaugh, 2007; Lappe, Herholz, Trainor, & Pantev, 2008). There is also evidence that music training can benefit non-music cognitive functions. For example, elderly musicians have enhanced cognitive and perceptual functions compared with non-musicians (Hanna-Pladdy & Mackay 2011; Zendel & Allain, 2012). Omigie and Samson (2014) have proposed that music training and expertise may be associated with greater "brain reserve capacity." This notion has preliminary support from the observation that playing a musical instrument is one activity that can reduce the risk of dementia (Verghese et al., 2003; Grant & Brody 2004). Although causal inferences are not possible, it is generally believed that musical instrument playing has concrete benefits for the aging brain.

Given the vast array of complex skills required in playing a musical instrument, it is remarkable that people with dementia can maintain and even learn this ability. In the case of persistent playing or new learning of a musical instrument after the onset of dementia, we propose that this activity has significant benefits for accessing memory and the self. Before addressing this hypothesis, we will discuss current published cases of preserved musical instrument playing in people with dementia.

## Musical Instrument Playing in People With Dementia

Musical instrument playing has been reported in people with two types of dementia, namely AD and FTD (various subtypes). Table 8.1 summarizes 11 published case studies to date, with findings of neuroimaging investigations and assessments of music and non-music memory functions. The majority of these cases have described individuals with AD (7/11, and an additional one that had a differential diagnosis of Primary Progressive Aphasia, PPA), with the remaining cases being variants of FTD (PPA, Semantic Dementia - SD, or Bv-FTD; see Table 8.1). Of note, only one diagnosis was confirmed at autopsy (Beatty, Brumback, & Vonsattel, 1997), whereas the remainder were clinical diagnoses based on neuroimaging results and cognitive assessments, typically the Mini Mental State Examination (MMSE), or in some cases the Addenbrooke's Cognitive Examination (ACE) or more detailed neuropsychological assessments. The majority (8/11) of published cases are male, and the most common instrument is the piano (7/11). In 5 cases, the ability to play a musical instrument persisted despite impairments associated with a severe stage of dementia according to MMSE scores (Beatty et al., 1988; Cho et al., 2015; Fornazzari et al., 2006; Polk & Kertesz, 1993).

In four cases, neuroimaging investigations revealed greater left hemisphere damage, suggesting a relative sparing of right hemisphere brain functions (Cho et al., 2015; Fornazzari et al., 2006; Polk & Kertesz, 1993; Weinstein et al., 2001). In the AD cases, one showed no abnormalities, while five cases showed diffuse cortical atrophy and three showed temporal lobe

**Table 8.1.** A Summary of Nine Published and Four New Cases of Musical Instrument Playing in People With Dementia

| Authors (year) | Sex/ Age | Dementia Type | Neuroimaging | Severity of Dementia | Instrument/Music Training | Music Memory Explicit | Music Memory Implicit | Non-Music Memory Explicit | Non-Music Memory Implicit |
|---|---|---|---|---|---|---|---|---|---|
| 1. Beatty et al. (1988) | F/81 | AD | MRI: diffuse cortical atrophy EEG: diffuse slowing | MMSE 17/30 then 8/30 10 months later. Severe anomia, reduced verbal fluency, receptive aphasia, constructional apraxia, and ideomotor apraxia. | Piano/master degree in music, taught music at college level and private piano lessons. | Semantic: variable | Familiar songs with lyrics but no music. Played a simple song on an alternative instrument (xylophone). Sight read a new song published after dementia onset. | Impaired remote memory (famous faces and Fargo map test). | Impaired (no learning on the PRT or GPT). |
| 2. Crystal et al. (1989) | M/82 | AD | CT and EEG both normal. | Annual cognitive assessment for 7 years. IQ remained stable until Year 6, stable performance of Raven Progressive Matrices | Piano Musicologist with 12 years formal training, 40 years as a music editor. Played 2 hours daily during adulthood. | Semantic: impaired. | Able to continue playing popular classical pieces after the first few bars were played. | Impaired verbal memory (Fuld Object Memory Evaluation). | Intact (improved speed reading mirror reversed words). |
| 3. Polk & Kertesz (1993) | M/58 | PPA or AD | CT and MRI showed enlarged ventricles and diffuse cerebral atrophy with greater left involvement. | MMSE 3/30 | Guitar/music teacher with 12 years education. | Semantic: variable | Able to improvise and play fluently (although noted to be perseverative and jargonistic). | Not assessed due to severe aphasia. | N/A |

*(continued)*

**Table 8.1.** Continued

| Authors (year) | Sex/ Age | Dementia Type | Neuroimaging | Severity of Dementia | Instrument/Music Training | Music Memory Explicit | Music Memory Implicit | Non-Music Memory Explicit | Non-Music Memory Implicit |
|---|---|---|---|---|---|---|---|---|---|
| 4. Beatty et al. (1994) | M/71 | AD | MRI showed cortical atrophy (marked in temporo-parietal region) and a possible lacuna in the left Globus pallidus and substantial atrophy of the cerebellar hemispheres. AD confirmed by pathology.* | MMSE 20/30 | Trombone/ amateur musician who played in a Dixieland Jazz band. | Semantic: reduced relative to controls | Continued to play in Dixieland jazz band without music. No significant difference in ratings of performances pre- and post-onset of dementia | Impaired verbal and visual memory (Recognition Memory test for words and faces) and remote memory (famous faces). | Intact (Normal learning on PRT)** |
| 5. Weinstein et al. (2001) | M/64 | SemD | MRI left anterior temporal atrophy | MMSE 27/30, 23/30, 14/30 over 3 annual reviews. Impaired performance of language tasks but intact visual perception. | Piano, organ, harpsichord performer | No formal assessment but noted that his embellishments during playing reflect his "meaningful and productive control over musical knowledge." | Performed familiar Baroque pieces with embellishments and sight read unfamiliar pieces. | N/A | N/A |

| Study | Sex/Age | Diagnosis | Neuroimaging | Cognitive | Musical background | Music assessment | Music memory | Memory | Other |
|---|---|---|---|---|---|---|---|---|---|
| 6. Cowles et al. (2003) | M/80 | AD | MRI: bi-temporal atrophy with prominent sulci, ventricles, and cisterns. Small infarcts of left basal ganglia and thalamus, punctate infarct of the right cerebellar hemisphere. Periventricular white matter changes. | MMSE 16/30 Normal visual perception, moderate attention and mild language deficits. | Violin/private piano and violin lessons during school, 1 year of music training at conservatorium. Claimed he played 32 instruments. Music teaching and amateur performance. | Semantic: mildly impaired on identification of Christmas Tunes but intact Seashore Rhythm test. | Able to learn and recall an unfamiliar piece published post dementia onset after a 10-minute delay. | Impaired verbal, visual memory (RBANS) and remote memory (autobiographical and famous aces). Impaired performance of environmental sounds and musical instrument memory tests. | Intact (Normal learning on PRT and some priming evident on GPT). |
| 7. Fornazzari et al. (2006) | F/63 | AD | MRI: diffuse mild cerebral atrophy; SPECT: asymmetric hypoperfusion in left frontotemporal and parietal regions. | MMSE 10/30, 1 year later 5/30. | Piano/ professional | No formal assessment but able to play familiar music and read and interpret unfamiliar music. | Able to learn and recall a new piece (in auditory but not written form) presented daily over 1 week. Able to play familiar music. | Impaired verbal memory (auditory and written, specific test not reported). | N/A |
| 8. Hsieh et al. (2009) | M/70 | Bv-FTD | MRI: significant orbitofrontal and anterior temporal atrophy. | ACE-R 70/100, 1 year later 50/100. Intact visuospatial skills. | Piano/18 years education, piano lessons since childhood. | No formal assessment but played familiar music. | Played familiar piece with music. No difference in ratings of two filmed performances 1 year apart. | Reduced memory (verbal) score on ACE-R. | N/A |

(continued)

**Table 8.1.** Continued

| Authors (year) | Sex/Age | Dementia Type | Neuroimaging | Severity of Dementia | Instrument/Music Training | Music Memory | | Non-Music Memory | |
| --- | --- | --- | --- | --- | --- | --- | --- | --- | --- |
| | | | | | | Explicit | Implicit | Explicit | Implicit |
| 9. Cho et al. (2015) | M/59 | Bv-FTD | MRI: bilateral anterior temporal and dorsolateral and medial frontal atrophy, more marked on left. PET: hypo-metabolism in bilateral dorsolateral and medial frontal and anterior temporal lobe, more severe on left side. | MMSE 26/30, 2 years later 7/30 | Saxophone/no formal music training and never played an instrument training pre-dementia onset. Only learned post dementia. Practiced 2 hours daily for 3 years. Represented his music class in a contest. By Year 4 was unable to learn new songs but continued to play previously learned songs by sight reading. | No formal assessment but teacher reported that he had "advanced skills in accuracy, rhythm and tempo and was a good sight reader." | Started saxophone lessons at time of diagnosis, 2 hours daily. Learned to read music and played unfamiliar songs. | Impaired verbal memory (SVLT) but intact visual memory (RCF) at Year 1, impaired verbal and visual memory at Year 3. | N/A |

| 10. Fornazzari et al. (2017) | Case 1: M/87 Case 2: F/82 | Case 1 and 2 Probable AD | Case 1 CT: moderate to severe generalized atrophy. Case 2: CT diffuse generalized atrophy. | Case 1: MMSE 13/30; MoCA 10/30. Case 2: MMSE 19/30; MoCA 19/30. | Case 1: Piano, 10 years formal training, ongoing regular playing. Case 2: Piano, 5 years formal lessons in high school. | Not formally assessed but both able to sight read novel piece. Case 1: aware of musical errors, and when improvising, maintained the key of the piece, showed knowledge of musical parameters. | Learned a novel piece over 7 days. Both cases showed improvement in their delayed recall of the piece. Case 1: recalled 4 bars with both hands by the 7th training day. Case 2: recalled 4 bars with right hand on 7th day. | Impaired verbal recall learning and memory (logical memory—short story recall from WMS). | N/A |

---

ACE-R = Addenbrooke's Cognitive Evaluation-Revised; ACE-3 Addenbrooke's Cognitive Evaluation-Third version; AD = Alzheimer's Dementia; Bv-FTD = Behavioural Variant Frontotemporal dementia; CT = Computed tomography; EEG = electroencephalogram; F = female; GPT = Gollin Picture Test; M = male; MRI = Magnetic resonance imaging; PET = positron emission tomography; PPA = Primary Progressive Aphasia, PRT = Pursuit Rotor Test; SemD = Semantic dementia; SPECT = single photon emission computed tomography; MMSE = mini-mental state examination, MoCA = Montreal Cognitive Assessment; RBANS = Repeatable Battery for the Assessment of Neuropsychological Status; RCF = Rey Complex Figure; SVLT = Shiraz Verbal Learning Test; WMS: Wechsler Memory Scale.

*Beatty et al. (1997)

** but noted that he had dressing apraxia and was unable to tie his own necktie atrophy. In the two cases of Bv-FTD, one showed orbitofrontal and anterior temporal atrophy (Hsieh, Mioshi, Baker, Piguet, & Hodges, 2009) and the other had atrophy and hypometabolism (shown on MRI and PET respectively) of bilateral dorsolateral, medial frontal, and anterior temporal regions, more marked on the left (Cho et al., 2015).

Although many skills of music performance reflect procedural, embodied memory, the integrity of semantic music memory can also be assessed by musical instrument playing. For example, when a musician responds to a request to play a piece that is named rather than demonstrated, or when a musician identifies elements of music structure, such as a note or a chord. Semantic music memory was reported to be impaired or variable in the cases in which it was formally assessed (Beatty et al., 1988, 1994; Cowles et al., 2003; Crystal et al., 1989; Polk & Kertesz, 1993; see Table 8.1). There were methodological differences between the studies in how semantic music memory was assessed, such as recalling the title or composer of the music (Beatty et al., 1988, 1994; Crystal et al., 1989; Polk & Kertesz, 1993), or recalling the last note by singing (Polk & Kertesz, 1993). For example, in Beatty et al.'s (1988) case of an 81-year-old piano player with AD, they found some aspects of her music knowledge were intact (e.g., playing scales to command, identification of clefs, major and minor keys, and sight reading and transposing of unfamiliar music), while others were impaired (e.g., inability to identify time or key signatures, inability to complete the Seashore rhythm test, and severe impairment in identification of titles of Christmas songs or classical works). Interestingly, she was able to play Christmas songs whose titles she could not recall or recognise. Similarly, Crystal et al. (1989) found that their 82-year-old piano player with AD was unable to recall the title or composers of famous classical pieces. He could continue to play classical pieces when cued by the first few bars but could not name or identify the titles or composer of any of the pieces he played. Polk & Kertesz's (1993) study of a guitarist with AD (with a differential diagnosis of PPA) showed that he could not read, write, or copy music notation. He could complete familiar melodic phrases by singing or whistling the last note, but could not name melodies. In summary, it appears that these cases typically showed *impaired* semantic music memory when assessed with language-based tasks requiring naming or identification of titles or composers of music, but *intact* semantic music memory when demonstrated by other (procedural) means, such as being able to play or sing familiar pieces when requested with the title. Behavioural-based measures of semantic music memory and familiarity decision tasks have revealed preservation of this type of memory in people with AD (see Cuddy & Duffin, 2005), as discussed in the section "Accessing Memory" below.

All the cases reported intact playing of familiar pieces or spared "procedural" music memory. This sparing of procedural music memory was demonstrated through cued recall (playing the first bar and asking the patient to continue; Crystal et al., 1989) or playing music from their own repertoire (see Table 8.1). In one case the person was able to play a simple piece ("Twinkle Twinkle Little Star") on a new instrument (xylophone; Beatty et al., 1988). Some cases were able to sight read and learn unfamiliar pieces or even learn and recall pieces published post onset of dementia (Cowles et al., 2003; Fornazzari et al., 2006; Fornazzari et al., 2017). Only the pianist with AD reported by Beatty and colleagues (1988) had healthy control data for comparison. They compared their patient's recorded performances of six songs with four neurologically intact pianists and had three independent raters rank the pianists. They described the patient's proficiency as "somewhere between that of the young, once proficient but somewhat rusty amateur player and elderly, once accomplished pianist whose current performance was impaired by arthritis" (Beatty et al., 1988, p. 161).

In two cases, longitudinal data in the form of recorded performances were obtained (Beatty et al., 1994; Hsieh et al., 2009). Beatty et al. (1994) described a trombone player with AD who was a longstanding Dixieland band member. They compared current (post-dementia) recordings of his performances in the band to those three decades earlier (pre-dementia). A professional jazz horn player rated the patient's "tone, intonation, overall quality of his ensemble, solo and style to be of good quality," and he noted no changes between the pre- and post-onset of AD recordings. The rater observed only a "slight decrease in technical skills manifest as a slightly lessened ability to play fast passages cleanly and a reduced tendency to play higher registers" (Beatty et al., 1994, p. 1042). One tune that was performed by the band on both occasions (pre- and post-dementia) was also rated by 35 neuropsychology students, professionals, and technicians. They found no significant difference in the average ratings of the pre- and post-dementia recordings, and 40% of raters indicated that they could not tell which version had been recorded after he developed dementia. The authors concluded that "it seems reasonable to conclude that the patient has not suffered any meaningful loss in his ability to play the trombone as a consequence of his dementia" (Beatty et al., 1994, p. 1043). In their study of a pianist with Bv-FTD, Hsieh et al. (2009) compared ratings (of note and rhythm accuracy, tone quality, dynamics, tempo, and interpretation) made by professional music teachers of two video recorded performances taken one year apart (both recordings post-dementia). Although the ratings declined, there were no statistically significant differences. These observations suggest that musical instrument playing can be a relatively stable skill in the face of cognitive impairments associated with AD or Bv-FTD.

All cases showed impaired performances of non-music explicit memory tasks, which typically assessed verbal memory using a variety of tasks (see Table 8.1). With regard to implicit non-music memory, only 4/11 cases formally examined this, and all but one (Beatty et al., 1988) showed intact implicit learning. Most studies (3/4) used the Gollin Picture Completion or Pursuit Rotor tests (Beatty et al., 1988; 1994; Cowles et al., 2003), and one used reading of mirror writing (Crystal et al., 1989, see Table 8.1). The Gollin Picture Completion task comprises degraded line drawings of common objects that become progressively more complete over three trials, and the complete object is presented at the third trial. Subjects are required to name the object at each trial. The Pursuit Rotor Test requires the subject to maintain contact of a stylus on a circular spot on a spinning turntable.

Beatty et al.'s (1988) case of musical instrument playing in the context of impaired performance on these implicit non-musical memory tasks suggests that procedural music memory, or the ability to play a musical instrument, is not dependent on the type of implicit memory functions that are measured by the Gollin Picture (visual) or Pursuit Rotor (motor) tasks. In general, implicit memory function is regarded to be relatively intact in people with AD, but it is noteworthy that most studies have used motor skill learning tasks (see review by van Halteren-van Tilborg, Scherder, & Hulstijn, 2007). Studies of people with AD using the Gollin Picture Completion task have typically found impaired performance relative to healthy elderly (Corkin, 1982; Mack et al., 1993). Further research is required to characterise the nature of implicit memory function using different assessment tasks across motor and sensory domains (visual, auditory) in people with different types of dementia, and more specifically in those who play a musical instrument.

## Neural Correlates of Spared Music Skills in AD and FTD

Enhanced artistic skills, including music skills, have been observed in people with Bv-FTD. Miller and colleagues (2000) described two people who developed new musical skills after the onset of their dementia, namely whistling and composing, and another person who engaged in persistent piano playing throughout the progression of his dementia. All three of these patients showed left anterior temporal dysfunction with relative sparing of frontal regions on neuroimaging investigations (Single Photon Emission Computed Tomography, SPECT, or Positron Emission Tomography, PET). The authors drew on Kapur's (1996) theory of "paradoxical functional facilitation," and hypothesized that "selective degeneration of the left anterior temporal cortex led to decreased inhibition of the more right sided and posteriorly located visual and music systems, and dorsolateral systems involved with working memory, thereby enhancing artistic interest and productivity" (Miller et al., 2000, p. 462)

Insight into the neurological basis of preserved musical skill in dementia can also be gained from studies of other types of spared music skills, namely "musicophilia" or dramatically increased engagement in music in people with FTD, and preserved memory for familiar music and music-evoked autobiographical memories (MEAMs) in people with AD.

Recent research has documented 'musicophilia' or a strong desire to engage in music listening or other forms of music engagement in people with FTD. The phenomenon may involve music craving or seeking, with some patients "demanding to listen to a narrow repertoire of songs for up to many hours each day but sometimes also engaging in more organized behaviors such as taking up a musical instrument or buying music equipment" (Fletcher et al., 2015). This symptom was reported by carers of nearly half (48%) of Fletcher et al.'s (2015) sample of 56 people with various types of FTD and was more frequently observed in people with SD compared with Bv-FTD. In their initial study, musicophilia was associated with relative *preservation* of the volume of the left posterior hippocampus (Fletcher, Downey, Witoonpanich, & Warren, 2013), which is a distinct explanation to Miller et al.'s (2000) hypothesis. In Fletcher et al.'s (2015) subsequent study, however, no neuroanatomical associations were identified at the prescribed significance threshold, but a post hoc analysis at a more lenient threshold ($p <$ .001 uncorrected over the whole brain) revealed relative preservation of grey matter in the right hippocampus, which is more in keeping with the hypothesis of Miller et al. (2000).

There is increasing evidence of spared memory for familiar music in people with AD, which is strikingly contrasted by their impaired memory for verbal and visual material. The ability for familiar music to trigger personal or autobiographical memories (music-evoked autobiographical memories or MEAMs) can also be preserved in people with AD (Baird, Brancatisano, Gelding, & Thompson, 2018; Cuddy, Sikka, Silveira, Bai, & Vanstone, 2017; El Haj, Fasotti, & Allain, 2012). Various characteristics of MEAMs of people with AD are in keeping with healthy people, including the frequency (Baird et al., 2018; Cuddy et al., 2017), length (word count) of the reported memories, and vividness ratings (Cuddy et al., 2017).

Among those with AD, there is also no significant relationship between frequency of MEAMS and the severity of cognitive impairment (Baird et al., 2018; Cuddy et al., 2017). This observation provides further support for the notion that music abilities, including music memory, often survive neurodegeneration compared with other nonmusic skills. This apparent resistance may reflect the widespread brain regions that are activated by music, as demonstrated by accumulating functional neuroimaging research (e.g., Koelsch, 2011). Furthermore, several neuroimaging studies have identified a crucial role of the medial frontal regions in MEAM retrieval (Ford, Addis, & Giovanello, 2011; Ford, Rubin, & Giovanello, 2016; Janata, 2009). In the case of preserved MEAMs in people with AD, it has been shown that the frontal brain regions are relatively less affected by AD pathology (Jacobsen et al., 2015).

Jacobsen and colleagues (2015) combined Magnetic Resonance Imaging (MRI) and PET to compare neural responses to three types of music excerpts that were unknown, recently learned, or long known in healthy young people. They found that frontal brain regions, specifically the caudal anterior cingulate and ventral presupplementary motor area, were crucial for long-term music memory. In a group of 20 people with AD, they found that these brain regions were less affected by AD pathology compared with the rest of the brain, in that they showed less cortical atrophy (on MRI) and minimal disruption of glucose metabolism (on PET). Nevertheless, amyloid-beta deposition was not lower in the music memory region of interest compared with the rest of the brain. Amyloid beta deposition is the result of a disruption of balance between the production and clearance of amyloid precursor protein, which leads to the formation of amyloid-beta plaques, development of neurofibrillary tangles, neural dysfunction, regional atrophy, and finally dementia, also known as the "amyloid cascade hypothesis" (Hardy & Higgins, 1992). Amyloid-beta deposition is expected to precede cortical hypometabolism and atrophy. Jacobsen and colleagues (2015) concluded that the music memory brain region of interest is in the very early stages of AD biomarker development, but relatively spared of hypometabolism and cortical atrophy, which explains the remarkable preservation of music memory in people with AD.

As discussed, neuroimaging studies of healthy people have shown that the medial prefrontal cortex (which includes the anterior cingulate cortex, identified as part of the "music memory region of interest" by Jacobsen et al, 2015) plays a crucial role in the retrieval of MEAMs (Ford, Addis, & Giovanello, 2011; Ford, Rubin, & Giovanello, 2016; Janata, 2009). The medial prefrontal cortex is part of the Default Mode Network (DMN), which is activated during self-related functions including self-referential thoughts and self-awareness (see Northoff et al., 2006 for a review). We have proposed that activation of this brain region during listening to personally preferred music may underlie the self-enhancing effect of music in both healthy people and those with dementia (Baird & Thompson, 2018). We now extend this notion to musical instrument playing, and suggest that this frontal brain region, critical for both retrieval of MEAMs and self-related processing, is also engaged during musical practice.

The pattern of AD pathology and related neural dysfunction with relative sparing of frontal brain regions could explain why music can activate powerful memories and a sense of self in this population. In the case of FTD, however, patterns of pathology are more variable and depend on the subtype. Nevertheless, we can hypothesize that there may be relative preservation of medial frontal regions in those people with FTD who show preserved or post dementia onset de novo musical instrument playing.

In the next section we discuss our proposal that playing a musical instrument in the face of dementia allows a unique form of access to memory and the self, and we address each of these domains in turn.

## Accessing Memory

We propose that playing a musical instrument facilitates retrieval of certain types of memories (in both music and non-music domains) that can remain spared in people with AD and Bv-FTD. In the musical domain, procedural, semantic, and episodic music memories can be stimulated through musical instrument playing. Autobiographical or personal memories associated with previous (or future) music playing can also be activated. These personal memories can then stimulate related autobiographical memories of significant people or events in life, which may be strongly linked with the sense of self. In this way, music performance provides a unique form of "access" to memory and the self, which may not be possible through verbal or visual stimuli. How does music playing achieve this? We propose that the mechanisms underlying this unique access to memory and the self are related to both the specific design features of music (Thompson & Schlaug, 2015) and the widespread neural correlates engaged by music playing, including medial frontal regions. We will now discuss the different forms of music (semantic, procedural, and episodic music memory) and non-music (autobiographical) memory that are accessed by music instrument playing in people with dementia.

Impaired explicit memory function is the hallmark symptom of AD, but implicit (in particular procedural motor skills learning) has been found to be relatively intact. In accordance with the different forms of memory proposed by Tulving and his colleagues (e.g., Schacter & Tulving, 1994; Tulving, 1985, 1995), it follows that various types of musical memory exist and may be differentially spared and impaired in people with AD. In general, implicit musical memory is expected to remain preserved for longer than explicit musical memory during progression of the disease, but evidence for the preservation of both forms of musical memory has been reported. There are mixed results on the preservation of explicit music memory for familiar or unfamiliar music. A review of findings by Baird and Samson (2009) suggested that explicit memory for music—as assessed by yes/no recognition tasks, recalling the title/composer of music, or singing the rest of a musical phrase—is typically impaired in people with AD relative to healthy controls. Evidence of the sparing of musical memory in individuals with AD, however, was reported in the seminal case study of EN by Cuddy and Duffin (2005). Selected group studies also suggest that semantic musical memory can remain relatively preserved in AD (Vanstone et al., 2012; see also Cuddy, Sikka, & Vanstone, 2015). As discussed in the previous section, these inconsistent results may be explained by methodological differences in how semantic music memory was assessed, with better performance when behavioural measures or familiarity-based decision tasks are used compared with yes/no recognition tasks.

In the published cases in which formal assessment of semantic musical memory was conducted, several cases showed a dissociation between the recall or identification of the titles/composers of famous pieces, and production of such pieces through playing, singing, or whistling (Beatty et al., 1988; Crystal et al., 1989; Polk & Kertesz, 1993). Furthermore, all published cases were able to perform familiar pieces (see Table 8.1). These observations provide support for our proposal that the act of playing a musical instrument enables access to preserved forms of semantic music memory which may not accessed by other means.

Rare observations of the learning and recall of new music in people with AD provides further support for our notion that actively playing a musical instrument can access preserved forms of memory in the face of impaired learning and recall of non-music (e.g., verbal) material. New learning and recall of unfamiliar music has been reported in three pianists (Fornazzari et al., 2006; Fornazzari et al., 2017) and a violinist (Cowles et al., 2003). These observations suggest that episodic music memory can remain relatively preserved compared with episodic verbal and visual memory. This phenomenon is not restricted to those who play an instrument, as spared ability to learn new music by singing has also been observed in a non-musician with severe AD (Baird, Umbach, & Thompson, 2017). There are also remarkable cases of de novo learning of the saxophone and ukulele post onset of Bv-FTD (Cho et al., 2015; and our new Case 4 in Table 8.2). In one case, non-music recall memory was impaired (Cho et al., 2015), while in our case (Case 4 in Table 8.2) it was relatively preserved. Further studies comparing non-music and music memory in people with dementia, particularly non-AD types, are required to understand the relative patterns of integrity and impairment of these functions (see Chapter 5 for a discussion of music cognition in non-AD dementias).

Music is also uniquely effective at eliciting autobiographical memories and, as described above, MEAMs are preserved in people with AD. Previous studies of MEAMs have excluded musicians and have been restricted to people with limited music training. Cuddy et al. (2015, p. 227) reported that musicians "tend to produce memories of the musical or technical structure of pieces rather than personal memories." We are unaware of any formal published research on MEAMs in musicians (either healthy or with dementia). In our four new cases of musical instrument playing in people with dementia outlined below, we found that their music playing stimulated memories not only of past (or future) musical performances, but also of

Table 8.2. A Summary of Four New Cases of Musical Instrument Playing in People With Dementia

| Case | Sex/Age | Dementia Type | Neuroimaging | Severity of Dementia | Instrument/Music Training | Music Memory | Non-Music Memory | | |
|---|---|---|---|---|---|---|---|---|---|
| 1 | F/77 | AD | Not available | ACE-3: 32/100: 6-month review: 28/100 12-month review: 26/100 | Piano/formal lessons from 6 years then worked as a piano teacher | Famous tunes: 10/10 MBEA (melody): 7/8, 7/8, 5/8: Rhythm: 4/8 | Played familiar pieces with music and able to sight read new piece. | ACE (memory): 5/26: Verbal learning: 1/12, 3/12, 4/12 | GPT: 0/10, 2/10, 2/10 |
| 2 | F/95 | AD | Not available | ACE-3: 66/100: 12-month review: 70/100 | Piano/formal lessons from 11 years and played in church throughout life | Famous tunes: 8/10 MBEA (melody): 3/8, 7/8, 4/8 Rhythm: 5/8 | Played familiar pieces with music. Able to sight read new piece. | ACE (memory): 8/26: Verbal learning: 5/12, 5/12, 9/12, 4/12 (delay) | GPT: 0/10, 4/10, 5/10 |
| 3 | M/91 | AD | None available | ACE-3: 61/100 | Chanter/played bagpipes and drums in pipe band from early 20s. | Famous tunes: 10/10 MBEA (melody): 5/8, 6/8, 7/8: Rhythm: 7/8 | Played familiar pieces without music and able to learn a new piece accompanied by son. | ACE (memory): 9/26: Verbal learning: 3/12, 1/12, 6/12, 3/12 (delay) | GPT: 1/10, 2/10, 5/10 |

| 4 | M/78 | Bv-FTD | Frontotemporal cortical atrophy | ACE-3: 87/ 100 | Ukulele (taught himself post dementia onset/ singing lessons and amateur performer since aged 20 | Famous tunes: 10/ 10: MBEA (melody): 6/ 8, 7/8, 5/ 8: Rhythm: 6/ 8 | Sang familiar songs and also learned new songs (both solo and in singing groups). Taught himself ukulele post onset dementia to accompany himself while singing. | ACE (memory): 23/ 26; Verbal learning: 5/12, 5/12, 5/12, 2/ 12 (delayed 5 minutes) | GPT:3/10, 6/ 10, 6/10 |

F=female, M=male, AD=Alzheimer's Dementia, Bv-FTD=Behavioural Variant Frontotemporal dementia.

GPT=Gollin Picture Completion Test, three learning trials; ACE=Addenbrooke's Cognitive Evaluation-Revised; MBEA=Montreal Battery of Evaluation of Amusia (melody subset only), Rhythm= Rhythm Discrimination Task (identifying if two rhythms are the same or different); Famous tunes = familiarity decision test requiring identifying familiar tunes (Hsieh et al., 2011); Verbal learning=One Story from Logical Memory of Wechsler Memory Scale IV related personal memories such as a particular time of life (childhood) or significant people (e.g., sharing music with family members). In this way, musical instrument playing accesses autobiographical memories that are not only music specific, but also those that are linked to an individual's self-identity.

## Accessing the Self/Selves

What happens to your sense of self when you develop dementia? Our sense of self is widely regarded to be informed by our autobiographical memories, and also by our recognition and understanding of significant people in our immediate environment. It is therefore assumed that if our memories and recognition of significant people and our environment are disrupted by the pathology of dementia, then our self-identity may be "lost" or, at the very least, unsettled. The complex relationship between autobiographical memory and the self in the context of AD has been addressed by several authors (e.g., El Haj, Antoine, Nandrino, & Kapogiannis, 2015; Mograbi, Brown, & Morris, 2009; Tippett, Prebble, & Addis, 2018). A detailed discussion of this topic is beyond the scope of this chapter, but, in brief, theories range from those that consider that a decline in autobiographical memory corresponds with a diminished sense of, or loss of self (El Haj et al., 2015), or with a failure to update self-knowledge resulting in a static or "petrified" self (Mograbi, Brown, & Morris, 2009), to a more nuanced and multifaceted theory suggesting that the integrity of narrative construction from semantic memory underpins the sophistication and certainty in beliefs about self-persistence over time ("diachronic unity"; Tippett et al., 2018). Overall, these "memory centric" theories fail to consider the importance of other "non-memory" aspects of self, such as extended (physical objects) and situated (environmental) aspects. As argued by Gallagher (2013), despite the loss of ability to recall one's past life that can occur in AD, self-identity may continue to be supported by other aspects of the self. "This is not to say that such changes do not result in a modulation of self-experience or self-identity, but rather, since self is not reducible to any one of these aspects, it is a modulation rather than a complete loss. . . . If someone lacks memory or a sense of agency, or perhaps lacks both, she continues as a self if there are a sufficient number of aspects still intact" (Gallagher, 2013, p. 4).

In their review of the qualitative and quantitative methods used to study the self in dementia, Cadell and Clare (2010) identified that research on this topic has been challenged by two interrelated issues: (1) the numerous definitions and models of the self that exist within and across disciplines, and (2) methodological difficulties in designing measures of the self/selves in this population. These authors stated that it was difficult to form firm conclusions due to methodological variations between investigations which addressed different components of the self and reflected various models and concepts of self. The majority of qualitative studies were cases or small samples and typically demonstrated that the self was preserved, while the quantitative studies suggested that some components deteriorated as the disease progresses. They highlighted the need for further studies to be based on the development of measures to assess various aspects of the self, and a clear theoretical framework of the self, and suggested Neisser's (1988) model as most appropriate.

## Music as a Self-Enhacer

We have proposed that music is uniquely effective at enhancing the sense of self in people with dementia because of its combination of features and ability to engage various aspects of self simultaneously (Baird & Thompson, 2018). Specifically, we proposed that specific attributes or design features of music (cf. Thompson & Schlaug, 2015) can modulate the self. For example, music is social, personal, and emotional. Specifically, it can be a social activity that encourages physical and verbal interactions and facilitates communication between participants. Familiar and favorite music can stimulate personal memories and emotional responses. Sharing memories and emotions associated with favorite songs can facilitate social relationships.

Each of these features of music engages with different aspects of the self, enhancing the feeling of unity and continuity within the self (Baird & Thompson, 2018). Our novel theoretical framework draws on and extends the work of several other authors, including Gallagher (2013), Elvers (2016), and MacDonald, Hargreaves, and Miell (2017). For the purpose of our model, we restricted our discussion to five aspects of self, as outlined by Neisser (1988). We used this model as a template for our framework, both to contain and focus our discussion, and in support of the suggestion that it is an appropriate theoretical framework for future research on the topic of the self in dementia (Caddell & Clare, 2010).

Neisser (1988) defined five selves, as follows:

(1) Ecological self, or the self as it is directly perceived (through processing of sensory information) with respect to the physical environment. In other words, the position and movements of our bodies in the environment.

(2) Interpersonal self, or the self engaged in personal interactions (two or more people).

(3) Extended self, or the self as it was in the past and what we expect it to be in the future, primarily based on memory.

(4) Private self, or personal experiences that are not available to others, such as dreams and beliefs.

(5) Conceptual self, or self concept—a concept of oneself as particular person, for example, social roles and social/cultural differences.

Neisser states that these "selves are not generally experienced as separate and distinct, because there is stimulus information to specify their cohesion. In cases where such information is less salient, the unity of the self is correspondingly weakened. But unified or not, all fives 'selves' are of fundamental importance" (1988, p. 36).Individuals with dementia may exhibit a unique pattern of impairment or integrity of the different aspects of self. As such, different forms of music treatment can be used to target specific aspects of self. The model highlights how music is an optimal stimulus for treatment or enhancement of the self in people with dementia, given its unique design features and ability to engage multiple cognitive processes and dimensions of the self simultaneously (Baird & Thompson, 2018). The model is applicable to people with or without musical training, and the unique effects of music on the self can be seen in either musicians or non-musicians. Nevertheless, the impact of music on the self in those who play a musical instrument may be particularly powerful. In the case of those who played a musical instrument prior to the onset of their dementia, continuing to play after their diagnosis and as their condition progresses may be a means (and in some cases, the only way) of accessing their sense of self. It may be one of the few remaining skills that they have in the midst of a loss of skills in other domains. Thus, it serves as a unique means of self-preservation, by accessing who they were before the onset of dementia and preserving this in the face of dementia.

We propose that an important brain mechanism underlying the access to the self/selves through musical instrument playing is the activation of the medial frontal cortex. This brain region is part of the DMN (as discussed above), which is considered to be integral to self-processing, and has also been identified as crucial in retrieval of MEAMs, and found to be relatively spared of AD pathology. The engagement of the medial frontal cortex during musical instrument playing stimulates self-reflection and autobiographical memories, providing a direct link to an individual's personal past and future. This hypothesis requires future investigation using neuroimaging techniques.

Rare cases in which people learn a musical instrument *post* dementia onset (see Cho et al. in Table 8.1 and Case 4 in Table 8.2), highlight the remarkable potential for the self to continue to evolve in the context of dementia. If a new musical skill is developed in a person who was already musically trained and engaged in music (see Case 4 below), it may represent a modulation or extension of the pre-dementia self/selves. Alternatively, in the case of a person who had no previous musical training or engagement in any musical activities prior to their dementia onset (as in the case reported by Cho et al., see Table 8.1), their post-dementia learning of a musical instrument may additionally be considered a "de novo" aspect of self, a "musical self," expressed through playing an instrument, representing a novel form of self-expression. In either case, musical instrument playing in people with dementia highlights that our self does not remain static and is certainly not "lost" in the face of a neurodegenerative condition. Rather, former self/selves can be modulated or extended, or a new aspect of self, a "musical self," can emerge in the face of dementia, and musical instrument playing is a unique means of expressing this.

We will now discuss four new cases of musical instrument playing in people with dementia. Their demographics and cognitive (music and non-music domains) assessment results are summarized in Table 8.2. There are three cases of probable AD (two female pianists, one male bagpipe player) and one male with probable Bv-FTD who was an amateur singer who taught himself the ukulele post onset of his dementia. All were able to play familiar pieces and sight read a new piece. In regard to semantic music memory, all showed preserved identification of famous tunes, but their melody and rhythm perception was variable (see Table 8.2). With the exception of Case 4 with Bv-FTD, all showed significantly impaired verbal learning and recall memory. All four cases showed evidence of implicit learning on the Gollin Picture Completion Test (see Table 8.2).

In the next section we will describe our four new cases and highlight how their musical instrument playing accesses their five selves (in italics) and the specific features of music involved (in brackets).

## Four New Cases of Musical Instrument Playing in Dementia

### Case 1: Betina

Case 1 is taken from Baird and Thomson (2018). Betina is a 77-year-old woman with severe AD who resides in an aged care facility. She had piano lessons from 6 years of age and worked as a piano teacher. She plays familiar repertoire for at least one hour a day and shows spared music cognition skills (such as ability to sight read new pieces, perception of pitch, rhythm, and emotions portrayed by music, completing melodies from an initial prompt and recognition of familiar melodies) in contrast with severe impairments in other cognitive domains (see Table 8.2).

When asked why she continues to play the piano, she said "I just like the music and also sometimes people can't play so I take over, even at a wedding nobody can play and I played." She is motivated to play well and commented "I always play good. If I can't play good I not feel good." When asked how she feels when she plays well, she responded "I feel happy. I can push the music to the people who come in here and they will feel quite good." The proposed impact of music on Betina's five selves is outlined below.

The physical demands of playing the piano require her to have a strong sense of her own body and its interaction with the environment, specifically the keyboard. These bodily sensations contribute to a strengthening of the *ecological self*. Piano playing requires fine motor control and this physical action would create a sense of embodiment (physical). Her *interpersonal self* is facilitated by music in that she enjoys playing for others (social, engaging) and has a special role as a pianist, playing on special occasions or for grace at meals. This gives her a sense of purpose, identity (*conceptual self*), and pride in her ability to make others "feel good" (social, engaging, persuasive). She is also able to play with others (synchrony), as evidenced by her ability to play piano duets with author AB. She appeared to enjoy this and requested to meet and play together again in the future. This highlights how music can enhance her *extended self*, by prompting her sense of the future (wanting to play again, practicing to play well), but also triggering autobiographical memories (personal, emotional) and discussions of her past (for example memories of her music lessons and performances during childhood). Her *private self* is expressed through her daily practice for at least one hour a day on her own in the chapel within the facility. She initiates this herself and it stimulates positive emotions. She says she feels "happy" when she plays and this is evident qualitatively in her demeanor after playing. Furthermore, she is able to complete musical phrases suggesting that her musical imagery skills remain preserved (personal, *private self*). Her *conceptual self* is evident in her preferences for specific music genres, namely familiar classical repertoire, her identity as a musician and the associated role she has within the facility. In summary, Betina engages with music on a daily basis through her regular practice and occasional performances for others. Her use of music and response to it demonstrates how the various features of music can facilitate all aspects of her self, and create an overall enhanced sense of self.

## Case 2: Bess

Bess is a 94-year-old woman living in a residential age care facility. She has not received a formal diagnosis of AD, but review of her medical records revealed that on her admission to the facility 4 years ago a geriatrician noted "cognitive impairment," and raised the question of this diagnosis. Her score on a cognitive screening tool (Addenbrooke's Cognitive Examination-III) was in keeping with a diagnosis of dementia (see Table 8.2), and with memory difficulties being the most prominent feature of her cognitive deficits, AD is considered the most likely differential. She learned piano and organ initially from her mother and then had formal lessons with a neighbor from 11 years of age. She did music theory and also learned violin from 15–17 years of age. She played organ and piano in church services from the age of 15 and always uses sheet music when playing. At the aged care facility she currently plays organ in the chapel at the weekly service and piano for grace at meals three times a day.

Bess' *ecological self* is engaged in the physical requirements of playing keyboards (physical). Playing for others in the chapel and at meal times accesses her *interpersonal self* (social). Her *conceptual self* is evident in that she is known for her role as a musician by other residents (engaging, persuasive), and her music preferences (hymns and classical pieces). Her *extended self* is accessed through her regular routine of playing for specific future events (e.g., the next chapel service) and her memories of playing at previous similar events and enjoying music in her family life (personal, emotional). Her *private self*, specifically her spiritual/religious beliefs, are accessed through her playing at religious events (grace at meals and chapel services). In summary, Bess' keyboard playing accesses all five aspects of her self, and it is this simultaneous access that makes it a unique form of self-enhancement.

## Case 3: Roy

Roy was a 91-year old man who resided in an aged care facility. He played drums (side, tenor, and bass) and bagpipes in a pipe band from his early 20s. He rehearsed for at least 3 hours once a week and also played in regular band performances. He did not have formal lessons, rather he learned while playing with the band. His music reading ability was limited and he learned to play "by ear." One of his three sons also played in the band. He ceased playing for over a decade, but after his wife died and he moved into the facility his son brought him a chanter (upper portion of the bagpipes) and they began playing together during his visits.

Roy's *ecological self* was engaged when playing the chanter through the required breathing techniques (physical). His *interpersonal self* was most evident in the facilitated relationship with his son with whom he played duets (synchrony, social). His son composed a new bagpipe tune and dedicated it to his father and taught him how to play it by playing in synchrony. Roy's *conceptual self* was demonstrated in his musical preferences (Scottish pipe music) and his identity as a previous band member and as someone who was still able to play a musical instrument, despite his other cognitive impairments. His *extended self* was evident in his fond reminiscing of band days (personal, emotional). When Roy died, his son showed video footage at his funeral that had been taken of them playing together. It was a powerful reminder of his father in his pre-dementia days, and of their newly formed bond while playing duets post-dementia. In summary, playing the chanter enabled access to aspects of his self, in particular his interpersonal self (enhanced relationship with son), which may not have been accessible in any other way

## Case 4: Barry

Barry is a 78-year-old man with Bv-FTD who is cared for by his wife in at home. He commenced singing lessons during his 20s and has enjoyed performing in musicals and amateur community signing groups all his life. He ran his own singing group that performed at retirement villages for 5 years. He cannot read music and learns songs by ear and reading lyrics. He continues to sing in a local community choir and still enjoys performing. He learned the ukulele after the onset of his dementia by attending group lessons and accompanies himself while singing.

Singing and playing the ukulele simultaneously are both physically demanding activities, requiring breathing, fine motor control, and coordination, engaging the *ecological self*. This physical action would create a sense of embodiment (physical). Barry's *interpersonal self* is engaged by his singing and playing ukulele in group settings and performing for others (social, engaging, synchrony). His role as a singer, choir member, and ukulele player are important features of his *conceptual self* and identity, both past and present. Access to his *extended self* is evidenced by his desire to learn a new skill (ukulele), his practice for future performances, and recalling of autobiographical memories (personal, emotional) of previous singing groups and performances he has participated in. His *private self* is expressed through his daily practice, and his ability to complete musical phrases suggests preserved musical imagery skills (personal, *private self*). In summary, Barry's post-dementia ukulele playing can be considered an extension of his pre-dementia selves which he is able to access very efficiently through this new skill.

## Conclusion

Musical instrument playing by people with dementia enables a unique form of access to different types of memory and self, which may not be accessible by any other means. Our review of published cases and four new cases demonstrates how different types of music memory (procedural, semantic, episodic) and autobiographical memories (both music and non-music related) can be relatively preserved and accessed through musical instrument playing. Using our model of the impact of music on the self in dementia (Baird & Thompson, 2018), we have highlighted specific aspects of self (according to Neisser's framework) that can be accessed during musical instrument playing.

We propose that an important neural mechanism underlying preserved musical instrument playing in people with dementia and its ability to access memory and self is the relative preservation and activation of frontal brain regions, specifically the medial prefrontal cortex. This brain region is integral in the retrieval of MEAMs and self-related processing. Musical instrument playing engages this critical brain region, within the wider neural network that mediates this activity, which enables access to preserved forms of music memory and facilitates retrieval of autobiographical memories, both within a musical context and more generally in life, providing a link between the current, past, and even the future self. Musical instrument playing enables both self preservation and expression, and highlights that rather than a "loss of self" in people with dementia, there is potential for continued development of the predementia self/selves, or for the manifestation of a post-dementia new aspect of self, a "musical self." This demonstrates the remarkable capacity for the ongoing evolution of the self throughout the progression of a neurodegenerative disorder.

# References

Baird, A., Brancatisano, O., Gelding, R., & Thompson, W. (2018). Characterisation of music and photograph evoked autobiographical memories in people with Alzheimer's Dementia, *Journal of Alzheimer's Disease, 66*(2), 693–706. doi:10.3233/JAD-180627

Baird, A., & Samson, S. (2009). Memory for music in Alzheimer's disease: Unforgettable? *Neuropsychological Review, 19,* 85–101. doi:10.1007/s11065-009-9085-2

Baird, A., & Thompson, W. F. (2018). The impact of music on the self in dementia. *Journal of Alzheimer's Disease, 61*(3),827–841. doi:10.3233/JAD-170737

Baird, A., Umbach, H., & Thompson, W. (2017). A non-musician with severe Alzheimer's Dementia learns a new song. *Neurocase, 23,* 36–40. doi:10.1080/13554794.2017.1287278

Beatty, W. W., Brumback, R. A., & Vonsattel, J. P., (1997). Autopsy-proven Alzheimer disease in a patient with dementia who retained musical skill in life. *Archives of Neurology, 54,* 1448.

Beatty, W. W., Winn, P., Adams, R. L., Allen, E. W., Wilson, D. A., Prince, J. R., . . . Littleford, D. (1994). Preserved cognitive skills in dementia of the Alzheimer type. *Archives of Neurology, 51,* 1040–1046.

Beatty, W. W., Zavadil, K. D., Bailly, R. C., Rixen, G. J., Zavadil, L. E., Farnham, N., & Fisher, L. (1988). Preserved musical skill in a severely demented patient. *International Journal of Clinical Neuropsychology, 10*(4), 158–164.

Bugos, J. A., Perlstein, W. M., McCrae, C. S., Brophy, T. S., & Bedenbaugh, P. H. (2007). Individualized piano instruction enhances executive functioning and working memory in older adults. *Aging & Mental Health, 11,* 464–471. doi:10.1080/13607860601086504

Caddell, L. S., & Clare, L. (2010). The impact of dementia on self and identity: A systematic review. *Clinical Psychology Review, 30,* 113–126. doi:10.1016/j.cpr.2009.10.003

Cho, H., Chin, J., Suh, M. K., Kim, H. J., Kim, Y. J., Ye, B. S., . . . Na, D. L. (2015). Postmorbid learning of saxophone playing in a patient with frontotemporal dementia. *Neurocase 21,* 767–772. doi:10.1080/13554794.2014.992915

Corkin, S. (1982) Some relationships between global amnesias and the memory impairments in Alzheimer's disease. In S. Corkin, K. L. Davis, J. H. Growdon, E. Usdin, & R. J. Wurtman (Eds.), *Alzheimer's disease: A report of progress in research* (pp. 149–164). New York, NY: Raven.

Cowles, A., Beatty, W. W., Nixon, S. J., Lutz, L. J., Paulk, J., Paulk, K., & Ross, E. D. (2003). Musical skill in dementia: A violinist presumed to have Alzheimer's disease learns to place a new song. *Neurocase 9,* 493–503.

Crystal, H. A., Grober, E., & Masur, D. (1989). Preservation of musical memory in Alzheimer's disease. *Journal of Neurology, Neurosurgery & Psychiatry, 52,* 1415–1416.

Cuddy, L. L., & Duffin, J. (2005). Music, memory, and Alzheimer's disease: Is music recognition spared in dementia, and how can it be assessed? *Medical Hypotheses, 64,* 229–235.

Cuddy, L. L., Sikka, R., Silveira, K., Bai, S., & Vanstone, A. (2017). Music-evoked autobiographical memories in Alzheimer's disease: Evidence for a positivity effect. *Cogent Psychology, 4,* 1277578. doi:10.1080/23311908.2016.1277578

Cuddy, L. L., Sikka, R., & Vanstone, A. (2015). Preservation of musical memory and engagement in healthy aging and Alzheimer's disease. *Annals of the New York Academy of Sciences, 1337*(1), 223–231. doi:10.1111/nyas.12617

El Haj, M., Antoine, P., Nandrino, J. L., & Kapogiannis, D. (2015). Autobiographical memory decline in Alzheimer's disease, a theoretical and clinical overview. *Ageing Research Reviews, 23*(Pt B), 183e192. http://dx.doi.org/10.1016/j.arr.2015.07.001

El Haj, M., Fasotti, L., & Allain, P. (2012). The involuntary nature of music-evoked autobiographical memories in Alzheimer's disease. *Consciousness & Cognition, 21.* 238–246. doi:10.1016/j.concog.2011.12.005

Elvers, P. (2016). Songs for the ego: Theorizing musical self-enhancement. *Frontiers in Psychology, 7*(2), 1–11. doi:10.3389/fpsyg.2016.00002

Fletcher, P. D., Downey, L. E., Golden, H. L., Clark, C. N., Slattery, C. F., Paterson, R. W., . . . Warren, J. D. (2015). Auditory hedonic phenotypes in dementia: A behavioural and neuroanatomical analysis. *Cortex, 67,* 95–105. doi:10.1016/j.cortex.2015.03.021.

Fletcher, P. D., Downey, L. E., Witoonpanich, P., & Warren, J. D. (2013). The brain basis of musicophilia: Evidence from frontotemporal lobar degeneration. *Frontiers in Psychology, 4,* 347–353. doi:10.3389/fpsyg.2013.00347

Ford, J. H., Addis, D. R., & Giovanello, K. S. (2011). Differential neural activity during search of specific and general autobiographical memories elicited by musical cues. *Neuropsychologia, 49*(9), 2514–2526.

Ford, J. H., Rubin, D. C., & Giovanello, K. S. (2016). The effects of song familiarity and age on phenomenological characteristics and neural recruitment during autobiographical memory retrieval, *Psychomusicology, 26,* 199–210.

Fornazzari, L., Castle, T., Nadkarni, S., Ambrose, M., Miranda, D., Apanasiewicz, N., & Phillips, F. (2006). Preservation of episodic musical memory in a pianist with Alzheimer disease. *Neurology, 66,* 610–611.

Fornazzari, L. Mansur, A., Ancua, K. O., Schweizer, T. A., & Fischer, C. E. (2017). Always in tune: The unforgettable memory for music in Alzheimer's disease. *Canadian Journal of Neurological Sciences,* 44, 209–211.

Gallagher, S. (2013). A pattern theory of self. *Frontiers in Human Neuroscience, 7,* 443. doi:10.3389/fnhum.2013.00443

Grant, M. D., & Brody, J. A. (2004). Musical experience and dementia. Hypothesis. *Aging Clinical and Experimental Research, 16*(5), 403–405.

Hanna-Pladdy, B., & MacKay, A. (2011). The relation between instrumental musical activity and cognitive aging. *Neuropsychology, 25,* 378–386.

Hardy, J., & Higgins, G. (1992). Alzheimer's disease: The amyloid cascade hypothesis. *Science, 256,* 184–185.

Herholz, S. C., & Zatorre, R. J. (2012). Musical training as a framework for brain plasticity: Behavior, function and structure. *Neuron, 76,* 486–502.

Hsieh, S., Hornberger, M., Piguet, O., & Hodges, J. R. (2011). Neural basis of music knowledge: evidence from the dementias. *Brain, 134*(9), 2523–2534.

Hsieh, S, Mioshi, E, Baker, F., Piguet, O., & Hodges, J. R. (2009, December 15–18). Piano playing skills in a patient with frontotemporal dementia: A longitudinal case study. In A. Williamon, S. Pretty, and R. Buck, (Eds.), *Proceedings of the International Symposium on Performance Science,* Auckland New Zealand (pp. 301–306).

Jacobsen, J.-H., Stelzer, J., Fritz, T. H., Chételat, G., La Joie, R., & Turner, R. (2015). Why musical memory can be preserved in advanced Alzheimer's disease. *Brain, 138*(8), 2438–2450. doi:10.1093/brain/awv135

Janata, P. (2009). The neural architecture of music-evoked autobiographical memories. *Cerebral Cortex, 19,* 2579–2594. doi:10.1093/cercor/bhp008

Kapur, N. (1996). Paradoxical functional facilitation in brain-behaviour research. A critical review. *Brain, 119,* 1775–1790.

Koelsch, S. (2011). Toward a neural basis of music perception—A review and updated model. *Frontiers in Psychology* 2, 110. doi:10.3389/fpsyg.2011.00110

Lappe, C., Herholz, S. C., Trainor, L. J., & Pantev, C. (2008). Cortical plasticity induced by short-term unimodal and multimodal musical training. *Journal of Neuroscience, 28*(39), 9632–9639. doi:10.1523/JNEUROSCI.2254-08.2008

MacDonald, R., Hargreaves, D. J., & Miell, D. (2017). *Handbook of musical identities.* Oxford, UK: Oxford University Press.

Mack, J. L., Petterson, M. A., Schnell, A. H., & Whitehouse, P. J. (1993). Performance of subjects with probable Alzheimer's disease and normal elderly controls on the Gollin incomplete pictures test. *Perceptual and Motor Skills, 77,* 951–969.

Matthews, S. (2015). Dementia and the power of music therapy. *Bioethics, 29,* 573–579. doi:10.1111/bioe.12148

Miller, B. L., Boone, K., Cummings, J. L., Read, S. L., & Mishkin, F. (2000). Functional correlates of musical and visual ability in frontotemporal dementia. *The British Journal of Psychiatry, 176,* 458–463. doi:10.1192/bjp.176.5.458

Mograbi, D. C., Brown, R. G., & Morris, R. G. (2009) Anosognosia in Alzheimer's disease—The petrified self. *Consciousness & Cognition, 18,* 989–1003. doi:10.1016/j.concog.2009.07.005

Neisser, U. (1988). Five kinds of self-knowledge. *Philosophical Psychology, 1,* 35–59.

Northoff, G., Heinzel, A., de Greck, M., Bermpohl, F., Dobrowolny, H., & Panksepp, J. (2006) Self-referential processing in our brain—A meta-analysis of imaging studies on the self. *Neuroimage, 31,* 440–457. doi:10.1016/j.neuroimage.2005.12.002

Omigie, D., & Samson, S. (2014). A protective effect of musical expertise on cognitive outcome following brain damage? *Neuropsychological Review, 24,* 445–460.

Palmer, C. (1997). Music performance. *Annual Review of Psychology, 48*(1), 115–138.

Polk, M., & Kertesz, A. (1993). Music and language in degenerative disease of the brain. *Brain & Cognition, 22,* 98–117.

Schacter, D. L., & Tulving, E. (1994). What are the memory systems of 1994? In D. L. Schacter & E. Tulving (Eds.), *Memory systems* (pp. 1–38). Cambridge, MA: MIT Press.

Thompson, W. F., & Schlaug, G. (2015). The healing power of music. *Scientific American Mind, 26,* 33–41.

Tippett, L. J., Prebble, S. C., & Addis, D. R. (2018). The Persistence of the Self over Time in Mild Cognitive Impairment and Alzheimer's Disease. *Frontiers in Psychology, 9,* 94.

Tulving, E. (1985). How many memory systems are there? *American Psychologist, 40*(4), 385–398. doi:10.1037/0003-066X.40.4.385

Tulving, E. (1995). Organization of memory: Quo vadis? In M. S. Gazzaniga (Ed.), *The Cognitive Neurosciences* (pp. 839–847). Cambridge, MA: MIT Press

Van Halteren-van Tilborg, I. A. D. A., Scherder, E. J. A., & Hulstijn, W. (2007). Motor-skill learning in Alzheimer's disease: A review with an eye to the clinical practice. *Neuropsychology Review, 17*(3), 203–212. doi:10.1007/s11065-007-9030-1

Vanstone, A. D., Sikka, R., Tangness, L., Sham, R., Garcia, A., & Cuddy, L. L. (2012). Episodic and semantic memory for melodies in Alzheimer's disease. *Music Perception: An Interdisciplinary Journal, 29*(5), 501–507.

Verghese, J., Lipton, R. A., Katz, M. J., Hall, C. A., Derby, C. A., Kuslansky, G., . . . Buschke, H. (2003). Leisure activities and the risk of dementia in the elderly. *New England Journal of Medicine, 348*(25), 2508–2516.

Weinstein, J., Koenig, P., Gunawardena, D., McMillan, C., Bonner, M., & Grossman, M. (2011). Preserved musical semantic memory in semantic dementia. *Archives of Neurology, 68,* 248–250.

Zendel, B. R., & Alain, C. (2012). Musicians experience less age-related decline in central auditory processing. *Psychology & Aging, 27,* 410–417.

# PART III
# MUSIC THERAPY IN DEMENTIA CARE

# 9

# Approaches to Measuring the Impact of Music Therapy and Music Activities on People Living With Dementia

*Becky Dowson and Orii McDermott*

## Overview

This chapter aims to highlight some of the challenges involved in measuring the potential impact of music on people with dementia. We will explore key issues including the rigor of measures, the potential burden imposed by the data collection process, and the ethical issues around involving people with dementia in research. We will then provide an overview of commonly used outcome measures covering the five domains: behavioral and psychological symptoms of dementia, quality of life, physiological changes, cognitive function, and music-related behaviors.

Challenges around quantifying highly personal musical experiences are also discussed, with examples from music therapy case studies to illustrate how the outcomes that are deemed important in the therapy process might translate to outcome measures. We recommend that insights obtained from qualitative studies be integrated into designing and conducting a quantitative or a mixed-method study, so that we can develop a deeper understanding of the mechanisms by which people with dementia may benefit from music-based interventions.

## Introduction

The potential impact of music for people with dementia is beginning to be widely recognized. Viral social media clips of Henry, from the documentary film "Alive Inside" (Alive Inside Foundation, 2012), and Teddy, the

"Songaminute Man" (Songaminute Man, 2016), have brought music and dementia into the public consciousness in a vivid and engaging, though anecdotal, manner. There is also a growing understanding among health professionals and policy makers that the arts may contribute to health and well-being in previously unexplored ways (All-Party Parliamentary Group on Arts Health and Wellbeing, 2017; Skingley, Bungay, & Clift, 2011). It is in this context that practitioners, researchers, care providers and others may identify a need to collect and analyze data about the impact of music for people with dementia in a way which might allow us to measure its benefits and understand how we might maximize these.

Measuring the impact of music in dementia care may be required in a variety of contexts. For example, a music therapist might wish to take a more objective look at her own practice with people with dementia in order to further develop her skills. Perhaps a care home manager wants to know whether the music-based training they have implemented for staff in the home has had any effects on the residents' well-being, in order to decide whether the training is cost-effective. On a larger scale, researchers might design a randomized controlled trial to compare depressive symptoms in people with dementia who attend a weekly singing group versus those who do not, with the findings potentially being used to inform policy decisions made by government or charities. In other words, different skills, resources, and methods will be needed to obtain results that are appropriate to specific situations.

But how do we decide what to measure? Again, this depends on the context of our enquiries. At the most basic level, we need to have a clear idea of what we want to find out and the resources which are available. We also need to know to whom we are reporting the results, and thus how rigorous the findings need to be. We should have an awareness of the challenges involved in evaluating music interventions, the unavoidable limitations of such evaluations, and the impact they may have on the people involved. The following section introduces the reader to some of these challenges and to other considerations which researchers and evaluators in the field of music and dementia will need to take into account.

## Choosing Variables and Outcome Measures

Inevitably, the biggest question for someone who wishes to assess the effects of music is to consider what changes might be expected or observed. This decision could be based on previous research but it might also be informed by personal experience of how music affects people with dementia in a particular situation. However, the decision about what outcome measures to use will also be affected by a number of other factors. The next section provides an introduction to some of the considerations and challenges involved when deciding what variable(s) to investigate and how to do so.

## Rigor of Measures

Before data collection begins, a research question or a hypothesis should guide the choice of outcome measure(s) and study design. For example, a therapist might be assessing results from their own practice or monitoring the progress of particular clients. In this case, one option is to develop a scale for measuring particular outcomes, which can be tailored to specific clinical situations and include only categories which are directly relevant to the line of inquiry. The disadvantage of using a researcher-designed scale without psychometric evaluations is that the results yielded cannot be directly compared to other study results using different scales, and even to studies using the same scale if test-retest reliability has not been established. Someone conducting a service evaluation, or investigating how a new music intervention has affected service users, will have different priorities. In this case, it will not be practical to measure every single variable which might be affected by the intervention, and the enquiry may be based around the changes that are important to the stakeholders.

On a larger scale, researchers conducting an experimental trial of the effectiveness of an intervention will need to ensure that the measures they use have been subjected to a full assessment of their psychometric properties, including measures of reliability and validity. Using validated measures allows results to be compared with results from other studies which have measured the same variable. Although there may be clinical reasons for using nonvalidated measures, in a research context their use may affect the validity of results. The existing literature can be used to inform choice of a measure, such as whether it has been previously used and validated for people with dementia and whether there are any specific challenges associated with its use. In some cases, it may also be necessary to re-validate previously validated scales for use in a new context; for example, if a scale which has been validated for use with older people is to be used with a population of people with dementia, or if it is translated into a different language.

## Resources Required for Outcome Measurement in Dementia

Administering outcome measures can be time consuming, especially if several variables are measured at the same time. Some measures might have 40 or more questions, but others are much shorter, so the level of detail required is a factor in the choice of measure. Learning to administer the measures correctly may require some form of training, and the time and expense of this also needs to be considered. The resources available will depend very much on the context of the work: a dedicated research team will be able to accomplish more than a small number of people who are conducting the study alongside other duties. Asking staff who are already overstretched to add completing outcome measures to their workload may not be realistic. Expert help may be required in some cases; for example, a neuropsychologist may be required to conduct cognitive assessments, or a statistician may be needed to carry out statistical analyses.

## Burden Imposed by Outcome Measurement

Depending on the type and method of data collection, study participants may find outcome measures tiring, boring, repetitive, confusing, or upsetting. It is important to consider the potential burden imposed by measures and ways to minimize burden. People with dementia may find questions challenging to answer and be acutely aware of the gaps in their memory and knowledge. There is evidence that people with dementia can find measures of cognitive performance (such as the Mini Mental State Examination) distressing, which may affect their willingness to take part in further testing (Lai, Hawkins, Gross, & Karlawish, 2008). However, seeking information directly from people with dementia is desirable if it is possible, and there is evidence that people with moderate dementia can complete measures (e.g., of quality of life) without a decrease in reliability and validity as compared to people with mild dementia and healthy participants (Logsdon et al., 1999). If self-report is not possible and a proxy (such as a professional or family caregiver) is asked to respond instead, the outcome measurement may prove to be burdensome if the proxy is short of time and/or balancing caring duties. Observational measures may remove burden from people with dementia, but can be more time consuming and burdensome for the person who conducts these observational measures. This could be a caregiver, the person conducting the research, or an independent observer who is "blind" to the purpose of the study. An additional consideration is that anyone conducting observations may need to be trained in the methods used, demanding further time and resources.

## Who Is the Audience?

It may also be helpful to think about who is going to see the results of a study or evaluation when considering what data to collect and the best way to present it. In some contexts, tables of figures, graphs and statistical analyses are appropriate, but other audiences might prefer qualitative data, which tells a story. One approach is to present the detailed statistical data in full to the audience, as well as a summary using clear and simple language for the lay reader. Alternatively, a mixed-methods approach might be appropriate, in which qualitative and quantitative data are both collected and then synthesized.

## Challenges of Outcome Measurement in Dementia Care

There are certain challenges inherent in measuring the impact of music interventions on people with dementia. Most musical activities and therapies can be called "complex interventions," meaning that they have many interacting components (Craig, 2012). There are many different types of dementia, and every person is affected in different ways and will respond to music differently as well. Musical tastes, personality, and past experiences all play a role in determining how someone will respond to a musical stimulus. Applying an objective measurement instrument to a subjective experience may seem a crude way of capturing the multi-faceted phenomenon that is musical engagement, but it offers the opportunity to see if parts of that experience are shared by groups of people.

The question of who should respond to verbally based outcome measures is also a challenge. Cognitive impairment means that people with dementia may find it difficult to respond to questions because they cannot under-stand them, cannot remember relevant information, cannot communicate their answers, or become anxious in a "testing" situation. For this reason, it is common to use proxy respondents, where someone who knows the person with dementia well answers on their behalf. This approach may work well when the questions are factual, but if they deal with feelings and emo-tional responses they are less reliable. Logsdon et al. (1999) showed that in a sample of 77 people with dementia, only five were unable to complete the Quality of Life in Alzheimer's Disease measure. There is a strong case to be made for empowering people with dementia to respond on their own be-half as much as possible, but creative strategies may be needed to enable this to happen (such as using video footage to prompt memory). Furthermore, it is widely acknowledged in psychosocial research that the views of people with dementia do not always correlate with those of family and caregivers, so researchers should be mindful of this and exercise caution when using and interpreting proxy measures (Harmer & Orrell, 2008; Hoe, Katona, Orrell, & Livingston, 2007). Observational measures offer an alternative to self or proxy report, but these can be time consuming to conduct. For example, Dementia Care Mapping (Bradford Dementia Group, 1997) was developed as a means of evaluating the quality of dementia care but has also been used as an outcome measure for research. A trained mapper observes a person with dementia for a considerable period of time (usually at least 6 hours) and records their behavior and state of well- or ill-being using pre-defined codes. The result is a detailed picture of an individual's quality of life, but one which is time and resource intensive to produce.

Finally, it may be challenging to find an appropriate person to administer the outcome measures. An independent research assistant would minimize the risk of bias. If a practitioner administers outcome measures about their own work, there is a chance that the respondent may feel obliged to provide positive responses or not feel able to answer freely for fear of offending. On the other hand, some respondents may feel more comfortable answering questions with a familiar person than a stranger. Practitioners should be aware that introducing measurement or evaluation into their work may alter the dynamics of their relationship with the people they are working with.

## Quantitative Outcome Measures Used in Studies of Music and Dementia

We will now consider some of the different quantitative outcome measures which researchers have employed in studies of music in dementia. It is important to bear in mind that types of music-based interventions offered to people with dementia are diverse. Evaluating these interventions and comparing the outcomes between the studies can be a complex task. Furthermore, differentiation between music-therapy studies and nonmusic-therapy studies is not always consistent; for example, some of the music-listening studies where the intervention is provided by care home staff are published as "music therapy" studies, and several systematic reviews on "music therapy for dementia" include both music-therapy studies and nonmusic-therapy studies. We do not suggest one intervention is better than the other. However, it is important to clarify the differences between the interventions so that the clinical benefits of each for people with dementia can be optimized. In the following section we describe some of the most common quantitative outcome measures that have been used in published studies to measure the impact of music interventions for people with dementia. Care has been taken to distinguish between the different kinds of intervention which are discussed; these include music therapy, group singing, listening to live and recorded music, and interactive music sessions. It is worth remembering that even within studies of similar interventions there is a lot of scope for variation. For example, listening to recorded music could mean a personalized playlist or background music played during a meal. Therefore it should not be assumed that if two studies use similar-sounding interventions and the same outcome measure, the results will be directly comparable.

A further dimension that must be considered is that although many different diseases cause dementia, the majority of music and dementia studies either focus on Alzheimer's disease or do not distinguish between different types of dementia. There are several possible explanations for this trend; one is that it is simply reflective of a broader pattern in dementia research, perhaps because of the prevalence of Alzheimer's disease and its presence among public consciousness. Another possibility is that researchers in this field have felt the symptoms of dementia to be more relevant to their enquiries than the type of dementia. Developing our understanding of how people with different types of dementia may respond differently to music interventions may be a potential avenue for future research, and it may be relevant to produce new outcome measures for this reason.

The outcome measures which are used in music and dementia research are not usually developed specifically for this purpose. Most have been created for research purposes, although some may have started out as screening instruments or clinical tools and been adopted as outcome measures. Others may have been developed for broader populations, such as older people in general, rather than specifically people with dementia. A paper by Moniz-Cooke et al. (2008) and an update (Joint Programme Neurodegenerative Disease, 2015) aimed to establish a consensus on the recommended outcome measures for psychosocial interventions with people with dementia. Many of the measures cited in this chapter are discussed in these papers. The outcome measures mentioned in this chapter have been chosen because they were the most frequently used measures in a review of music and dementia studies (Dowson et al., in preparation). They are divided into five categories of outcome measurement: BPSD, quality of life, physiological symptoms, cognitive function, and music-related outcomes. It is common for research studies to use a number of different measures. Including more than one outcome measure allows researchers to compare the results from two measures of the same variable and to explore associations between changes in different variables. For example, one could investigate whether there is a correlation between changes in BPSD and quality of life. Table 9.1 summarizes the most commonly used outcome measures and the studies that use them. These studies and their findings are discussed in more detail in the following sections.

## Evaluation of Behavioral and Psychological Symptoms of Dementia

Outcome measures for Behavioral and Psychological Symptoms of Dementia (BPSD) may provide an overview of all symptoms or may focus on specific symptoms (such as agitation, anxiety, or disruptive vocalization, for example). The Neuro-Psychiatric Inventory (NPI) is a widely used tool for assessing BPSD across 10 or 12 domains (the 12-item version includes assessment of sleeping and eating behavior; Cummings et al., 1994). The caregiver of the person with dementia is the respondent, and screening questions are used to minimize the time spent answering. For example, a caregiver will

**Table 9.1.** Commonly Used Outcome Measures in Studies of Music and Dementia

| Outcome Measure | Description | Examples of Studies Using the Measure |
| --- | --- | --- |
| Neuropsychiatric Inventory (Cummings et al., 1994) | Caregiver rates frequency and severity of behavioral symptoms across 10 or 12 domains. NPI-NH is a version specifically for residential care. | Hsu et al. (2015); Narme et al. (2014); Raglio et al. (2015); Camic et al. (2011) |
| Cohen-Mansfield Agitation Inventory (Cohen-Mansfield et al., 1989) | Caregiver rates frequency and disruptiveness of 29 agitated behaviors. | Ridder et al. (2013); Vink et al. (2013b); Ho et al. (2011); Park & Pringle Specht (2009) |
| Cornell Scale for Depression in Dementia (Alexopoulos et al., 1998) | 19-item clinician-administered scale, including responses from a primary caregiver and a brief interview with the person living with dementia. | Ashida (2000); Chu et al. (2014); Clarkson et al. (2007); Raglio et al. (2015); Ray & Mittelman (2017) |
| Geriatric Depression Scale (Yesavage et al., 1982) | 15- and 30-item versions available, with yes/no questions. Can be used for self-report for people with dementia. | Ceccato et al. (2012); Choi et al. (2009); Guetin et al. (2009) |
| Quality of Life in Alzheimer's Disease (Logsdon et al., 1999) | 13 items are ranked on a 4-point scale as Poor, Fair, Good or Excellent. Can be completed by people with dementia, proxy, or both. | Davidson & Fedele (2011); Särkämö et al. (2014) |
| Cornell-Brown Scale for Quality of Life in Dementia (Ready et al., 2002) | 19 bipolar items which are completed by a proxy respondent. | Särkämö et al. (2014); Raglio et al. (2015) |
| Dementia Care Mapping (Brooker & Surr, 2006) | Trained observer records behavior states of the person with dementia over a period of at least 6 hours. | Hsu et al. (2015) |
| Mini-Mental State Examination (Folstein et al., 1975) | Originally a clinical screening tool. 11 tasks which are scored out of 30, where a score of less than 25 indicates possible cognitive impairment. | Bruer et al. (2007); Chu et al. (2014); Sanchez et al. (2016) |
| Music in Dementia Assessment Scale (McDermott et al., 2015) | Five aspects of musical experience are rated on a visual analogue scale by caregiver and music therapist. | McDermott et al. (2015) |

be asked whether the person with dementia displays a certain symptom, and if they answer positively they will be asked to rate the frequency and severity of these behaviors, as well as the amount of distress they cause to the caregiver. There is also a version of this tool which is specifically designed for use with people with dementia in residential care (Wood et al., 2000). The NPI has been used in studies involving music therapy, music listening, and singing, with mixed findings. Hsu et al., (2015) found a reduction in the overall NPI score of care home residents with dementia after 5 months of individual music therapy, while Narme et al. (2014) observed comparable results after 4 weeks of a music listening-based activity. However, Raglio et al. (2015) measured no difference in NPI score between control and intervention groups in a study comparing music therapy and listening to music with usual care, and Camic et al. (2011) saw NPI scores increase over the course of a 10-week singing group.

A frequently used measure for agitation is the Cohen-Mansfield Agitation Inventory (Cohen-Mansfield, Werner, & Marx, 1989), which takes up to 20 minutes to administer. The inventory is completed by a primary caregiver and asks them to rate 29 agitated behaviors on a 7-point scale of frequency, thinking about how often they have occurred over the past two weeks. Respondents also indicate how disruptive the behavior is. The measure has been used widely in various studies of music and dementia, and reductions in CMAI score have been observed in studies of music therapy (Lin et al., 2011; Ray & Mittelman, 2017) and music listening (Ho et al., 2011; Narme et al., 2014). However, other studies have shown less promising results; for example, Vink et al. (2013) compared music therapy for people with dementia to general activities and found that although CMAI scores reduced in both groups, music therapy did not have a significantly greater effect. It is also common for researchers to modify CMAI to fit the purposes of their study; for instance, Ridder et al. (2013) asked caregivers to consider behavior frequency over the past week, rather than the standard 2-week period.

Other common psychological outcome variables are depression and anxiety. The Cornell Scale for Depression in Dementia is a 19-item clinician-administered scale, and includes responses from a primary caregiver and a brief interview with the person with dementia, both of which should be taken into account when scoring the items (Alexopoulos, Abrams, Young, & Shamoian, 1988). This measure is intended to take around 30 minutes to administer. Three studies which used CSDD to evaluate music therapy's effects on depression in people with dementia found significant reduction in depressive symptoms (Ashida, 2000; Chu et al., 2014; Ray & Mittelman, 2017). However, Clarkson, Cassidy, and Eskes (2007) found that listening to live music concerts did not have any effect on CSDD scores, and a study previously mentioned in the context of NPI also found no difference in CSDD score between control and intervention groups (Raglio et al., 2015).

The Geriatric Depression Scale is intended for use with elderly populations but not specifically people with dementia (Yesavage et al., 1982). It is suitable for self-report among people with dementia (depending on dementia severity) as the questions only require yes/no answers, rather than Likert scale ratings which may be conceptually more difficult to understand. It exists in 15- and 30-item versions, allowing the researcher to choose according to the resources available and the level of detail required. A study of music therapy which used GDS showed evidence that improvement in symptoms of depression was maintained for up to 2 months after the cessation of therapy (Guetin et al., 2009), while two other music-therapy studies did not detect a significant change using the same instrument (Ceccato et al., 2012; Choi, Lee, Cheong, & Lee, 2009).

## Evaluation of Quality of Life in People With Dementia

Quality of life has not been quantitatively examined as an outcome variable as frequently as measures of BPSD; however, several measures of quality of life have been employed in the research literature. The Quality of Life in Alzheimer's Disease (QoL-AD) Scale is suitable for use with people with all types of dementia (Logsdon, Gibbons, McCurry, & Teri, 1999). The scale consists of 13 items, which are ranked on a 4-point scale as poor, fair, good, or excellent. It is administered to the person with dementia, a proxy respondent or both; the proxy and self-rated scores can be combined in a way which gives more weight to the self-rated score. It has been used in a study of group singing in which an overall change in QoL-AD score was not detected (Davidson & Fedele, 2011). Another instrument for measuring quality of life is the Cornell-Brown Scale for Quality of Life in Dementia (CBS-QoL) which consists of 19 bipolar items (for example, both anxiety symptoms and signs that the person with dementia feels comfortable and relaxed are assessed; Ready, Ott, Grace, & Fernandez, 2002). The CBS-QoL is administered to a proxy respondent. In a study which compared the effects of singing and listening to music, Särkämö et al., (2014) administered both CBS-QoL and

QoL-AD and detected a significant positive change in both measures for the music listening group. Raglio et al. (2015) also used the CBS-QoL in their study of the effects of music therapy and music listening in people with dementia, but did not detect any difference between control and intervention groups.

Dementia Care Mapping (Brooker & Surr, 2006) offers a different approach to measuring quality of life in people with dementia by using observation (Copeland, Abou-Saleh, & Blazer, 2010). It is time-intensive to administer, since a trained observer must record the behavior and state of ill-being or well-being of a person with dementia every 5 minutes over a period of 6 hours. However, it offers a means of assessing quality of life from the point of view of the person with dementia, which may be especially useful in more advanced dementia where self-report is not possible. In their study of music therapy in a residential dementia care setting, Hsu et al., (2015) found positive results where DCM scores for the intervention group increased over time, while control group scores decreased.

## Evaluation of Cognitive Function in People With Dementia

A number of studies have attempted to assess whether music interventions can effect any short- or long-term change in cognitive performance or whether they may delay cognitive deterioration. A commonly used measure is the Mini-Mental State Examination (MMSE) which was originally developed as a clinical screening tool but has become popular in research (Folstein, Folstein, & McHugh, 1975). It consists of 11 tasks, scored out of 30, where a score of less than 25 indicates the possible presence of cognitive impairment. The MMSE takes only 5–10 minutes to administer, which may account for its widespread use. It has been used in a number of music-therapy studies: Bruer, Spitznagel, and Cloninger (2007) found that MMSE scores of participants with dementia were significantly improved the day after a music-therapy session but not the following week, while Chu et al. (2014) saw MMSE scores increase after group music therapy, although this effect was limited to people with mild to moderate dementia. Sanchez et al. (2016) used a version of the MMSE which had been adapted for people with severe dementia in a study comparing music listening with multi-sensory stimulation, but found no difference between control and intervention groups.

There have also been a number of studies which investigate whether music can improve performance on a specific cognitive function, such as short- and long-term memory, visuo-spatial ability or word fluency. In these studies, the control group is often a group of people of comparable age who do not have dementia. For example, El Haj, Postal, and Philippe (2012) investigated recall of autobiographic memories after exposure to different music conditions and found that recall was highest in the favorite music condition for people with Alzheimer's disease, while healthy controls showed no difference in recall. Simmons-Stern, Budson, and Ally (2010) found that people with Alzheimer's disease recalled unfamiliar words more successfully when they were sung rather than spoken.

## Evaluation of Physiological Changes in People With Dementia

Several studies have attempted to evaluate whether music can cause physiological changes in people with a dementia diagnosis, either through biological markers or changes in observed behavior. Hormone samples may be collected from saliva and are markers for a wide range of variables. Suzuki et al. (2007) measured salivary chromogranin A and immunoglobulin A levels, biomarkers of stress and immune system function, in a study investigating the effects of long-term group music therapy. They found evidence suggesting that attending a music-therapy session could lead to lower levels of chromogranin A, although no changes in immunoglobulin A were observed. Valdiglesias et al. (2017) also measured chromogranin A levels in people with dementia who attended individualized music sessions or multisensory stimulation, but they did not observe any changes in this marker. It is important to consider the ethical aspects and clinical implications when collecting any type of physiological data since it may be experienced as physically invasive and distressing by participants with dementia, especially if they do not fully comprehend why it is happening.

Data relating to other physiological variables may be collected less invasively through observation; Gill and Englert (2013) investigated whether playing background music in a care home during mealtimes had an effect on number of falls for people with dementia. This can be seen as an example of one variable being used as a proxy for something else; a decrease in falls might be due to a decrease in agitation-induced walking around. Information on falls is recorded anyway, making data collection more straightforward. However, the researchers concluded that using music in this way did not have an effect on the number of falls. Also in the area of motor skills, Wittwer, Webster, and Hill (2013) showed that playing recorded music and metronome beats could improve the gait of people with Alzheimer's disease while they walked on a treadmill, a technique known as rhythmic auditory cueing.

Since malnutrition is a serious problem among people with dementia, several studies have investigated whether various types of musical intervention could improve nutrition by observing the proportion of a meal which was consumed: two of these indicated that playing recorded music during mealtimes had a positive effect on caloric intake (Thomas & Smith, 2009; Wong, Burford, Wyles, Mundy, & Sainsbury, 2008), while McHugh et al. (2012) found no effect of pre-meal "vocal re-creative music therapy" on food consumption during the subsequent meal.

## Assessment of Music-Related Behaviors

The outcome measures which have been discussed so far have focused on the impact of music interventions upon domains which are not music-related. A different approach is to investigate potential changes which may occur in response to music, as well as music-related behaviors. The Music in Dementia Assessment Scale was developed from an identified need for "a psychometrically validated music-therapy outcome measure that reflects a holistic picture of therapy outcomes including increased positive responses of people with dementia" (McDermott, Orrell, & Ridder, 2015, p. 234). The items for the scale were developed from interviews and focus groups which explored how people with dementia experienced music (McDermott, Orrell, & Ridder, 2014). The resultant tool uses visual analogue scales to rate five aspects of musical experience: interest, response, initiation, involvement and enjoyment. Raglio et al., (2015) analyzed the responses of people with dementia attending music-therapy sessions using a version of the "Music-therapy checklist," which assesses musical, verbal, and nonverbal behaviors during sessions. Results from the checklist showed an increase in communicative musical behaviors, a finding which the authors use to interpret changes in other quantitative measurements of neuro-psychiatric symptoms, quality of life, and mood. Ceccato et al. (2012) used an Italian tool, the Music Therapy Activity Scale (Scala di Valutazione dell' Attivita Musicoterapeutica) which evaluates the musical relationship between therapist and client, and found that the relationship improved significantly over the course of the study compared with the control group. The authors remark that it is surprising that no corresponding changes in agitation or depression were found; this perhaps highlights the difficulties in drawing links between assessment of musical aspects and other variables.

## The "Unmeasurable" and Qualitative Methods

A frequent issue encountered by researchers in the field of music and dementia is deciding what is meaningful to measure. So far, we have considered some of the options for quantitative outcome measurement and the issues involved in the selection of appropriate tools. We have also seen that the results from these studies are not always conclusive, that expected outcomes sometimes do not materialize, and that changes in variables do not necessarily correlate in predictable ways. In questioning why these results occur, we must consider the limitations of quantitative methods, and what the alternatives are when we are faced with the "unmeasurable." Many people will assert that music has inherent qualities which may effect powerful changes in dementia care, but capturing these changes via measurable outcomes is challenging and may also seem like a gross oversimplification of the process. Music therapists have often pointed out that quantitative measures can never fully capture the essence of the musical experience. If we conduct a study, choose a variable to measure, then examine our data and find no change, is it because there was nothing to detect, because we measured the wrong thing, or because the measure used was not sensitive enough to detect subtle changes? The question then becomes an ethical one: if we are convinced of the affordances of music to bring change, we must find sufficient and appropriate evidence to demonstrate this to others in, order to allow more people with dementia to access music therapy and music interventions.

So, the first question is: how can we best measure the effects of musical experiences for people with dementia? A second, related question concerns the nature of the outcome measures we choose and whether they reflect the outcomes which are important to the other stakeholders in the process. Clearly, if a study is designed to examine a particular outcome, somebody involved thinks that outcome is important—but who, and why? For example, "wandering" residents in a dementia care home may present a challenge to caregivers. A music intervention which has the effect of keeping people in their chairs for longer periods of time may be appreciated by the overstretched staff—but how do we know whether the residents themselves are any better off? Can we reconcile the outcomes which are important to healthcare professionals, to caregivers and family members, and, most crucially, to the people living with dementia themselves? For example, it is common for researchers to refer to agitation, anxiety and wandering as "Behavioral and Psychological Symptoms of Dementia"; however, dementia activist Kate Swaffer has campaigned against the use of the term, arguing that most so-called challenging behaviors are actually expressions of unmet needs rather than symptoms of disease (Swaffer, 2018). In the previous example, the success of the music intervention in decreasing wandering may be meeting the needs of the staff, rather than the people with dementia themselves.

If we want to explore the "unmeasurable" and develop our understanding of the experience of music, we may choose to use qualitative methodological approaches. These approaches have been applied in research into music and dementia, although not as extensively as quantitative methodologies. Qualitative research designs have addressed questions about the meaning and value of music to people with dementia, and the part it plays in their lives. Sixsmith and Gibson (2007) interviewed people with dementia and their caregivers to develop their understanding of what music meant to them and its effect on their well-being; they found that as well as being enjoyable and personally meaningful, music was also an important source of social connection and empowerment for people with dementia. McDermott et al. (2014) investigated the importance of music to people with dementia by conducting interviews and focus groups, and analyzed the data using the general inductive approach (Thomas, 2006) to develop a theoretical model of music in dementia. This model highlighted music's accessibility for people with dementia, its value in supporting their personal identity, and its potential to help build relationships. Götell et al. (2002) transcribed and analyzed the conversation between people with dementia and caregivers during morning care situations, with and without singing from the caregivers. They found that when caregivers sang, people with dementia seemed to implicitly understand what was happening and were less resistant to care. Osman, et al. (2014) conducted interviews with people with dementia and their caregivers to gain their perspective on the benefits of attending "Singing for the Brain" sessions. Analysis of the transcripts using thematic analysis (Braun & Clarke, 2006) revealed that participants felt the sessions improved relationships and helped them to accept and cope with their diagnosis. In all these studies, the choice of qualitative methodology is guided by the question that the researchers wished to answer. Some research questions may call for both quantitative and qualitative data and analyses.

Mixed-methods research aims to integrate qualitative and quantitative data and to adopt a pragmatic position which is driven by a research question rather than allegiance to a particular methodology (Tashakkori & Teddlie, 1998). The ability to combine and compare different kinds of data (a process known as triangulation) means the researcher can approach the same question from different angles. For example, if you wanted to find out whether a certain music intervention reduced symptoms of depression and how it did so, you might choose to use a quantitative outcome measure to detect any change in symptoms, and also interview the participants or their caregivers for insight into the mechanism through which the change might have occurred. Camic et al. (2011) used a mixed-methods approach to investigate whether attending a singing group would improve the quality of life of people with dementia. Although the qualitative data showed an extremely positive response to the group, they found that "standardised measures appeared less successful in detecting change over the course of this intervention for both people with dementia and family caregivers" (Camic et al., 2011, p. 171). This example highlights the potential difficulty of finding a suitable outcome measure even when benefits were reported by the participants.

## Case Study Examples of Outcome Assessment in Dementia Care

The final part of this chapter uses two music-therapy case studies involving people with dementia to encourage the reader to think about the changes brought about by therapy and whether these could be measured or not. Case study reports are common in music-therapy literature. They normally consist of a description of a piece of clinical work with a particular client, describing the reasons for referral, the course of the therapy, and its results. Very often, detailed reflection of the process from the therapist is also present. Case studies are often treated as anecdotal evidence, since they focus on a piece of work with a single person. However, given that they offer a detailed description of the therapy, complete with the reasons the therapy began and whether it fulfilled its aims, we can use case studies as a point of comparison for outcome measures used in studies. To what extent do case study examples reflect the widely used outcome measures in research? The following section presents two case studies taken from the literature, and explores the aims, processes, and outcomes of therapy, then compares these to the outcomes which have been discussed in the preceding part of this chapter.

**Case study example 1.** Paulette Kydd describes her work with Phil, an elderly man with a diagnosis of probable Alzheimer's disease (Kydd, 2001). Phil's professional life had been spent as a welder, and he was also a proficient player of the mandolin and banjo. He had recently been admitted to a care home, but had not adjusted well to life there. He was described as "depressed and reclusive" and "grumpy" (Kydd, 2001, p.105–106). He preferred to stay in his room and did not want to talk to other residents or join in with activities or outings. He overheard a music-therapy session and asked about it, and he was then invited to join the music-therapy group. He showed willingness to attend the sessions, but did not want to take part in some of the activities. He also repeatedly asked for the same songs, which could not always be accommodated within the structure of the group session. For this reason, the music therapist offered him individual sessions so that she could play the music he liked and make a recording of it for his personal use, so that he would not need to ask for the songs repeatedly in the group session.

Phil began to attend individual music-therapy sessions, during which he chose songs he would like to hear. The music therapist played these songs and encouraged him to join in, with his voice or on instruments. Kydd notes that Phil was reluctant to play instruments and observes that this is common in people with dementia who were formerly proficient musicians: "They seem to feel that if they cannot play an instrument or sing perfectly, they do not want to try at all" (Kydd, 2001, p. 106). Staff described a "more positive affect" in Phil since commencing music therapy, and he would ask after the music therapist and sometimes leave his room to go looking for her. He "seemed increasingly able to function in the group" (Kydd, 2001, p.106) and began to interact more with the people around him, paying attention, making eye contact and joking with staff and residents. Kydd also mentions that Phil showed evidence of improved memory as he was able to remember other residents' names and countries of origin.

Since Phil seemed proud of his career as a welder, the music therapist suggested writing a song about this experience. Together, they changed the words to an existing song: "All the nice girls love a sailor" and the music therapist made an audio recording of this song and others that he enjoyed. Phil's son brought in his old banjo, and although he was reluctant, the music therapist managed to coax him to play. After a few attempts he was able to play "Yankee Doodle" perfectly. Despite saying he "couldn't play anymore," his playing continued to improve. He began to initiate interactions in the group setting, suggested songs which fit the theme, and tried to make them fit if they didn't. He also played his banjo for the other residents and seemed pleased by their appreciation. In her summary of the results of therapy, Kydd states that the staff had commented that music therapy had helped Phil adjust to his new life situation and become more social, less confused, and happier. His family also visited more often because Phil was more pleasant to be around.

In this brief summary of a much more detailed case study, we see several outcomes of therapy which tie in with the quantitative measures that have already been discussed. Phil was initially thought to be depressed, but by the end of the course of individual therapy his depressive symptoms seemed to have been alleviated: he appeared happier and was looking forward to things. He also seemed to be less confused and showed improvements in memory; this may have been due to the improvement in his depression, as mood disorders can compound cognitive impairment (Neu et al., 2005). Phil's increased interest in socializing is harder to quantify, but it may be interpreted as perhaps contributing positively to his quality of life or well-being. More difficult to interpret is his appreciation of the personalized song, and his renewed interest in playing the banjo. Both may be seen as evidence of improvements in mood, but on a deeper level they seem to reflect Phil's psychological need to have his sense of identity reinforced, especially in the face of cognitive decline. At first, he felt that he had lost the ability to play; but with the right encouragement, he discovered that although his ability may not be what it was, he could still bring pleasure to himself and others though his music. This discovery may also have had implications for his self-esteem and sense of self-efficacy.

**Case study example 2.** Maria Radoje relates her work with Elsa, an elderly woman with a diagnosis of Alzheimer's disease who lived in the dementia unit of a care home (Radoje, 2014). Elsa was referred to music therapy by the multidisciplinary team at the unit to help her deal with the "challenging outbursts" she sometimes experienced. Elsa was Jewish, and she had escaped to the UK as a child at the start of the World War II. In the early stages of their work together, Radoje describes how Elsa talked about issues related to her traumatic childhood experiences and was preoccupied with asking "Where were you born?" Radoje reports her own fear that she "would be exposed as not being wholly British," and this fear was realized when Elsa spotted her surname on her name badge (2014, p. 27). Radoje realized that in working with Elsa she may have to explore resonances in her own family history.

The musical relationship between Elsa and her therapist was established through well-known songs of the 20s, 30s, and 40s, which they sang together. Radoje began to incorporate small percussion instruments into the music, and then they moved on to playing larger instruments together. Elsa seemed to form an attachment to the therapist. She felt safe enough to speak in German during the sessions, and Radoje wanted to help her affirm the Jewish part of her identity as she felt Elsa was rather culturally isolated within the unit. Radoje introduced traditional Jewish songs into the work, and Elsa taught her some German words and phrases. In the next phase of their work, Elsa explored aspects of her family history which were traumatic and difficult, sometimes experiencing them as present events due to her dementia. Elsa's music therapy continued until her death, and Radoje writes that she "believed the work remained valuable in that it was one of the few avenues for Elsa to express her unique cultural heritage in a safe way" (2014, p. 32).

This summary overly simplifies the many complex layers of this case study. Elsa faces the same challenges that may affect many people with dementia, but with the added difficulty of a traumatic personal history. Music therapy seems hugely valuable to Elsa; we see its potential realized in helping Elsa form a relationship with the therapist, her responses to music, and her willingness to explore her past. But it is difficult to pinpoint any changes which could be measured directly by quantitative outcome measures discussed earlier, except perhaps the music-related ones. Radoje does not mention whether the original reason for referral, Elsa's challenging behavior, was helped by music therapy. The story of Elsa's music therapy may be linked to the work of Tom Kitwood, who suggested that BPSD are actually an expression of unmet psychological needs (Kitwood, 1997). Without music therapy, Elsa's need to have her sense of identity recognized and reinforced might have been unmet, which could in turn lead to decreased well-being and quality of life, and perhaps increased her behavioral problems.

The purpose of these two case study examples is to emphasize the highly personal nature of musical experiences and therapeutic music interventions. Phil and Elsa's music-therapy programs were unique to them; the same approach may not have worked for other people living with dementia. Although there were several elements in each case which may have been possible to measure quantitatively, we cannot hope to detect and record every outcome of their therapy using quantitative measures alone. Quantitative methods can be used to monitor desirable change which might occur as a result of music interventions, but they do not tell us much about the experiences underlying the change, or how change may occur.

## Conclusion

In this chapter we have discussed outcome measures commonly used for evaluating the impact of music interventions in dementia care and their application in the literature. We have also highlighted some of the complex issues which must be considered when undertaking research or evaluation in this field. It is a challenge to find outcome measures that are valid, acceptable to participants, and which measure outcomes that are relevant to real-world clinical situations, especially where the complexities of music and musical relationships are involved. However, a growing body of evidence suggests that it is possible to capture and describe the effects of music interventions and desirable to do so in a way that meets the established standards for evidence-based practice. In order to develop our understanding of the benefits of music for people with dementia, it will be vital that researchers not only use clinically relevant validated outcome measures, but also integrate clinical knowledge obtained from qualitative studies when designing and conducting a quantitative or mixed methods study.

## References

Alexopoulos, G. S., Abrams, R. C., Young, R. C., & Shamoian, C. A. (1988). Cornell Scale for Depression in Dementia, *23*, 271–284.

Alive Inside Foundation. [BeAliveInside] (3rd July 2012). *(ORIGINAL) ALIVE INSIDE clip of HENRY—YouTube*. [Video file]. Retrieved from https://www.youtube.com/watch?v=Hlm0Qd4mP-I

All-Party Parliamentary Group on Arts Health and Wellbeing. (2017). *Creative health: The arts for health and wellbeing*. London.

Ashida, S. (2000). The effect of reminiscence music therapy sessions on changes in depressive symptoms in elderly persons with dementia. *Journal of Music Therapy, 37*(3), 170–182.

Bradford Dementia Group. (1997). *Evaluating dementia care: The DCM Method.* Bradford, UK: University of Bradford.

Braun, V., & Clarke, V. (2006). Using thematic analysis in psychology. *Qualitative Research in Psychology, 3,* 77–101. https://doi.org/10.1191/1478088706qp063oa

Brooker, D. J., & Surr, C. (2006). Dementia Care Mapping (DCM): Initial validation of DCM 8 in UK field trials. *International Journal of Geriatric Psychiatry, 21*(11), 1018–1025. https://doi.org/10.1002/gps.1600

Bruer, R. A., Spitznagel, E., & Cloninger, C. R. (2007). The temporal limits of cognitive change from music therapy in elderly persons with dementia or dementia-like cognitive impairment: A randomized controlled trial. *Journal of Music Therapy, 44*(4), 308–328.

Camic, P. M., Williams, C. M., & Meeten, F. (2011). Does a "Singing Together Group" improve the quality of life of people with a dementia and their carers? A pilot evaluation study. *Dementia, 12*(2), 157–176. https://doi.org/10.1177/1471301211422761

Ceccato, E., Vigato, G., Bonetto, C., Bevilacqua, A., Pizziolo, P., Crociani, S., . . . Barchi, E. (2012). STAM protocol in dementia: A multicenter, single-blind, randomized, and controlled trial. *American Journal of Alzheimer's Disease and Other Dementias, 27*(5), 301–310.

Choi, A.-N., Lee, M. S., Cheong, K.-J., & Lee, J.-S. (2009). Effects of group music intervention on behavioral and psychological symptoms in patients with dementia: A pilot-controlled trial. *International Journal of Neuroscience, 119*(4), 471–481.

Chu, H., Yang, C.-Y., Lin, Y., Ou, K.-L., Lee, T.-Y., O'Brien, A. P., & Chou, K.-R. (2014). The impact of group music therapy on depression and cognition in elderly persons with dementia: A randomized controlled study. *Biological Research for Nursing, 16*(2), 209–217.

Clarkson, K. A., Cassidy, K.-L., & Eskes, G. A. (2007). Singing soothes: Music concerts for the management of agitation in older adults with dementia. *Canadian Journal of Geriatrics, 10*(3), 80–87.

Cohen-Mansfield, J., Werner, P., & Marx, M. S. (1989). An observational study of agitation in agitated nursing home residents. *International Psychogeriatrics, 1*(2), 153–165. https://doi.org/10.1017/S1041610289000165

Copeland, J., Abou-Saleh, M. T., & Blazer, D. G. (2010). *Principles and practice of geriatric psychiatry* (3rd ed.). Chichester: Wiley. https://onlinelibrary.wiley.com/doi/book/10.1002/9780470669600

Craig, P. (2012). *Developing and evaluating complex interventions.* Retrieved from www.mrc.ac.uk/complexinterventionsguidance

Cummings, J. L., Mega, M., Gray, K., Rosenberg-Thompson, S., Carusi, D. A., & Gornbein, J. (1994). The Neuropsychiatric Inventory: Comprehensive assessment of psychopathology in dementia. *Neurology, 44*(12), 2308–2308. https://doi.org/10.1212/WNL.44.12.2308

Davidson, J. W., & Fedele, J. (2011). Investigating group singing activity with people with dementia and their caregivers: Problems and positive prospects. *Musicae Scientiae, 15*(3), 402–422. https://doi.org/10.1177/1029864911410954

El Haj, M., Postal, V., & Philippe, A. (2012). Music enhances autobiographical memory in mild Alzheimer's disease. *Educational Gerontology, 38*, 30–41.

Folstein, M. F., Folstein, S. E., & McHugh, P. R. (1975). "Min.-mental state": A practical method for grading the cognitive state of patients for the clinician. *Journal of Psychiatric Research, 12*(3), 189–198. https://doi.org/10.1016/0022-3956(75)90026-6

Gill, L. M., & Englert, N. C. (2013). A music intervention's effect on falls in a dementia unit. *Journal for Nurse Practitioners, 9*(9), 562–567. https://doi.org/10.1016/j.nurpra.2013.05.005

Gotell, E., Brown, S., & Ekman, S.-L. (2002). Caregiver singing and background music in dementia care. *Western Journal of Nursing Research, 24*(2), 195–216.

Guetin, S., Portet, F., Picot, M.-C., Pommie, C., Messaoudi, M.. Djabelkir, L., . . . Touchon, J. (2009). Effect of music therapy on anxiety and depression in patients with Alzheimer's type dementia: Randomised, controlled study. *Dementia and Geriatric Cognitive Disorders, 28*(1), 36–46. https://doi.org/10.1159/000229024

Harmer, B. J., & Orrell, M. (2008). What is meaningful activity for people with dementia living in care homes? A comparison of the views of older people with dementia, staff, and family carers. *Aging & Mental Health, 12*(5), 548–558.

Ho, S.-Y., Lai, H.-L., Jeng, S.-Y., Tang, C.-W., Sung, H.-C., & Chen, P.-W. (2011). The Effects of researcher-composed music at mealtime on agitation in nursing home residents with dementia. *Archives of Psychiatric Nursing, 25*(6), e49–e55.

Hoe, J., Katona, C., Orrell, M., & Livingston, G. (2007). Quality of life in dementia: Care recipient and caregiver perceptions of quality of life in dementia: The LASER-AD study. *International Journal of Geriatric Psychiatry, 22*, 1031–1036. https://doi.org/10.1002/gps

Hsu, M. H., Flowerdew, R., Parker, M., Fachner, J. J., & Odell-Miller, H. (2015). Individual music therapy for managing neuropsychiatric symptoms for people with dementia and their carers: A cluster randomised controlled feasibility study. *BMC Geriatrics, 15*, 84. https://doi.org/10.1186/s12877-015-0082-4

Joint Programme Neurodegenerative Disease. (2015). *Dementia outcome measures: Charting new territory.* Retrieved from http://www.neurodegenerationresearch.eu/wp-content/uploads/2015/10/JPND-Report-Fountain.pdf

Kitwood, T. (1997). *Dementia reconsidered: The person comes first.* Maidenhead, UK: Open University Press.

Kydd, P. (2001). Using music therapy to help a client with Alzheimer's disease adapt to long-term care. *American Journal of Alzheimer's Disease and Other Dementias, 16*(2), 103–108.

Lai, J. M., Hawkins, K. A., Gross, C. P., & Karlawish, J. H. (2008). Self-reported distress after cognitive testing in patients with Alzheimer's disease. *The Journals of Gerontology: Series A: Biological Sciences and Medical Sciences, 63*(8), 855–859.

Lin, Y., Chu, H., Yang, C. Y., Chen, C. H., Chen, S. G., Chang, H. J., . . . Chou, K. R. (2011). Effectiveness of group music intervention against agitated behavior in elderly persons with dementia. *International Journal of Geriatric Psychiatry, 26*(7), 670–678. https://doi.org/10.1002/gps.2580

Logsdon, R. G., Gibbons, L. E., McCurry, S. M., & Teri, L. (1999). Quality of life in Alzheimer's disease: Patient and caregiver reports. *Journal of Mental Health and Aging, 5*(1), 21–32.

McDermott, O., Orrell, M., & Ridder, H. M. (2014). The importance of music for people with dementia: The perspectives of people with dementia, family carers, staff and music therapists. *Aging & Mental Health, 18*(6), 706–716. https://doi.org/10.1080/13607863.2013.875124

McDermott, O., Orrell, M., & Ridder, H. M. (2015). The development of Music in Dementia Assessment Scales (MiDAS). *Nordic Journal of Music Therapy, 24*(3), 232–251. https://doi.org/10.1080/08098131.2014.907333

McHugh, L., Gardstrom, S., Hiller, J., Brewer, M., & Diestelkamp, W. S. (2012). The effect of pre-meal, vocal re-creative music therapy on nutritional intake of residents with Alzheimer's disease and related dementias: A pilot study. *Music Therapy Perspectives, 30*(1), 32–42.

Moniz-Cook, E., Vernooij-Dassen, M., Woods, R., Verhey, F., Chattat, R., De Vugt, M., . . . Interdem, M. O. (2008). A European consensus on outcome measures for psychosocial intervention research in dementia care. *Aging & Mental Health, 12*(1), 14–29. https://doi.org/10.1080/13607860801919850

Narme, P., Clement, S., Ehrle, N., Schiaratura, L., Vachez, S., Courtaigne, B., . . . Samson, S. (2014). Efficacy of musical interventions in dementia: Evidence from a randomized controlled trial. *Journal of Alzheimer's Disease, 38*(2), 359–369.

Neu, P., Bajbouj, M., Schilling, A., Godemann, F., Berman, R. M., & Schlattmann, P. (2005). Cognitive function over the treatment course of depression in middle-aged patients: Correlation with brain MRI signal hyperintensities. *Journal of Psychiatric Research, 39*(2), 129–135. https://doi.org/10.1016/j.jpsychires.2004.06.004

Osman, S. E., Tischler, V., & Schneider, J. (2014). Singing for the Brain: A qualitative study exploring the health and well-being benefits of singing for people with dementia and their carers. *Dementia, 15*(6), 1326–1339. https://doi.org/10.1177/1471301214556291

Radoje, M. (2014). Where were you born? A music therapy case study. *British Journal of Music Therapy, 28*(2), 25–35. https://doi.org/10.1177/135945751402800206

Raglio, A., Bellandi, D., Baiardi, P., Gianotti, M., Ubezio, M. C., Zanacchi, E., . . . Stramba-Badiale, M. (2015). Effect of active music therapy and individualized listening to music on dementia: A multicenter randomized controlled trial. *Journal of the American Geriatrics Society, 63*(8), 1534–1539.

Ray, K. D., & Mittelman, M. S. (2017). Music therapy: A nonpharmacological approach to the care of agitation and depressive symptoms for nursing home residents with dementia. *Dementia (14713012), 16*(6), 689–710. https://doi.org/10.1177/1471301215613779

Ready, R. E., Ott, B. R., Grace, J., & Fernandez, I. (2002). The Cornell-Brown Scale for Quality of Life in dementia. *Alzheimer Disease and Associated Disorders, 16*(2), 109–115. https://doi.org/10.1097/00002093-200204000-00008

Ridder, H. M. O., Stige, B., Qvale, L. G., & Gold, C. (2013). Individual music therapy for agitation in dementia: An exploratory randomized controlled trial. *Aging & Mental Health, 17*(6), 667–678. https://doi.org/10.1080/13607863.2013.790926

Sanchez, A., Maseda, A., Marante-Moar, M. P., de Labra, C., & Lorenzo-Lopez, L. (2016). Comparing the effects of multisensory stimulation and individualized music sessions on elderly people with severe dementia: A randomized controlled trial. *Journal of Alzheimer's Disease, 52*(1), 303–315. https://doi.org/10.3233/JAD-151150

Särkämö, T., Tervaniemi, M., Laitinen, S., Numminen, A., Kurki, M., Johnson, J. K., & Rantanen, P. (2014). Cognitive, emotional, and social benefits of regular musical activities in early dementia: Randomized controlled study. *The Gerontologist, 54*(4), 634–650. https://doi.org/geront/gnt100

Simmons-Stern, N. R., Budson, A. E., & Ally, B. A. (2010). Music as a memory enhancer in patients with Alzheimer's disease. *Neuropsychologia, 48*(10), 3164–3167.

Sixsmith, A., & Gibson, G. (2007). Music and the wellbeing of people with dementia. *Ageing & Society, 27*(1), 127–145.

Skingley, A., Bungay, H., & Clift, S. (2011). Researching participatory arts, well-being and health: Some methodological issues. *Journal of Arts and Communities, 3*(1), 73–87. https://doi.org/10.1386/jaac.3.1.73

Songaminute Man. (4th August 2016). Quando quando quando. [Video file] Retrieved from https://www.youtube.com/watch?v=9UQ5mjFzHTA

Suzuki, M., Kanamori, M., Nagasawa, S., Tokiko, I., & Takayuki, S. (2007). Music therapy-induced changes in behavioral evaluations, and saliva chromogranin A and immunoglobulin A concentrations in elderly patients with senile dementia. *Geriatrics & Gerontology International, 7*(1), 61–71.

Swaffer, K. (2018, August 13). Rethinking dementia care: #BanBPSD [Web log post]. Retrieved from https://kateswaffer.com/2018/08/13/rethinking-dementia-care-banbpsd/

Tashakkori, A., & Teddlie, C. (1998). Mixed methodology Combining qualitative and quantitative approaches. In *Applied Social Research Methods Series* (Vol. 46, p. 183). Thousand Oaks, CA: Sage. https://us.sagepub.com/en-us/nam/mixed-methodology/book6245

Thomas, D. R. (2006). A general inductive approach for analyzing qualitative evaluation data. *American Journal of Evaluation, 27*(2), 237–246. https://doi.org/10.1177/1098214005283748

Thomas, D. W., & Smith, M. (2009). The effect of music on caloric consumption among nursing home residents with dementia of the Alzheimer's type. *Activities, Adaptation & Aging, 33,* 1–16.

Valdiglesias, V., Maseda, A., Lorenzo-Lopez, L., Pasaro, E., & Millan-Calenti. J. C. (2017). Is salivary chromogranin A a valid psychological stress biomarker during sensory stimulation in people with advanced dementia? *Journal of Alzheimer's Disease, 55*(4), 1509–1517. https://doi.org/10.3233/JAD-160893

Vink, A. C., Zuidersma, M., Boersma, F., De Jonge, P., Zuidema, S. U., & Slaets, J. P. J. (2013). The effect of music therapy compared with general recreational activities in reducing agitation in people with dementia: A randomised controlled trial. *International Journal of Geriatric Psychiatry, 28*(10), 1031–1038. https://doi.org/10.1002/gps.3924

Wittwer, J. E., Webster, K. E., & Hill, K. (2013). Effect of rhythmic auditory cueing on gait in people with Alzheimer disease. *Archives of Physical Medicine and Rehabilitation, 94*(4), 718–724.

Wong, A., Burford, S., Wyles, C. L., Mundy, H., & Sainsbury, R. (2008). Evaluation of strategies to improve nutrition in people with dementia in an assessment unit. *The Journal of Nutrition, Health & Aging, 12*(5), 309–312.

Wood, S., Cummings, J. L., Hsu, M. A., Barclay, T., Wheatley, M. V., Yarema, K. T., & Schnelle, J. F. (2000). The use of the neuropsychiatric inventory in nursing home residents: Characterization and measurement. *American Journal of Geriatric Psychiatry, 8*(1), 75–83. https://doi.org/10.1097/00019442-200002000-00010

Yesavage, J. A., Brink, T. L., Rose, T. L., Lum, O., Huang, V., Adey, M., & Leirer, V. O. (1982). Development and validation of a geriatric depression screening scale: A preliminary report. *Journal of Psychiatric Research, 17*(1), 37–49. https://doi.org/10.1016/0022-3956(82)90033-4

# 10

# Music and Music-Therapy Interventions for Behavioral and Psychological Symptoms of Dementia

## An Umbrella Review and Recommendations for Best Practice

*Anne W. Lipe and Molly Edmonston*

## Introduction

*Case Scenario:* Mrs. D resides in an assisted living memory care neighborhood and suffers from moderate dementia with episodes of anxiety and agitation. An activities assessment revealed that Mrs. D. used to enjoy dancing with her husband every Saturday night and that her favorite genre of music is big band swing. When Mrs. D starts experiencing symptoms of anxiety and agitation, she paces in and out of other residents' rooms and says, 'I need to find my children, they need to come home." When the activities coordinator observes these symptoms, she redirects Mrs. D to a quiet area of the neighborhood and begins to play big band swing music on an iPod. As the music plays, Mrs. D starts tapping her feet, nodding her head, and eventually wants to dance with the activities coordinator.

In the case scenario described above, it is evident that the music intervention helped to de-escalate Mrs. D's anxious and agitated behaviors and allowed her to transition to a calmer, happier state of being. Both pharmacological and nonpharmacological approaches, including music, have been used to address behavioral and psychological symptoms of dementia (BPSD) such as anxiety and agitation, but because side effects of medications often can be unpredictable, nonpharmacological approaches are preferred (Kar, 2009). The effectiveness of music and music therapy in addressing BPSD has been the topic of research for several decades, but diverse methodologies used across this research base make it difficult for practitioners to draw meaningful conclusions. A systematic review of the literature can be a useful way to synthesize findings from selected individual studies and to form meaningful conclusions about the direction of research efforts. However, systematic reviews have begun to proliferate, adding another layer of challenge to forming meaningful conclusions about the effectiveness of a treatment intervention. In this case, a review of reviews, sometimes called an overview of reviews or "umbrella review" (Aromataris et al., 2015, p. 133) may be conducted. The defining characteristic of the umbrella review is that its inclusion criteria are restricted to systematic reviews themselves rather than to individual studies. They are designed to summarize and compare important research findings for use in making decisions about best practice (Aromataris et al., 2015).

The purpose of the umbrella review conducted for this chapter was to summarize and compare the evidence in existing systematic reviews on the effectiveness of music and music-therapy interventions in addressing BPSD. Consistent with the outcomes of such an examination, recommendations for best practice are made. Of particular interest were (a) type of music interventions studied, (b) study designs used, (c) which BPSD were studied and how treatment outcomes were measured, (d) whether a credentialed music therapist provided the treatment intervention or was involved in study design and (e) the strength of results of meta-analyses.

## Overview to BPSD

While the identification of behavioral and psychological disturbances among individuals with dementia is not a new phenomenon, the term BPSD or "behavioral and psychological symptoms of dementia" entered the medical lexicon in 1996 when it was first used by the International Psychogeriatric Association (Milano, Saturnino, & Capasso, 2013). The term BPSD refers to a variety of neuropsychiatric symptoms characterized by disturbances in thought, emotional experience, perception, and motor function (Cerejeira, Lagarto, & Mukaetova-Ladinska, 2012). The most frequently occurring BPSD are depression, anxiety, agitation, irritability, and apathy (van der Linde, Stephan, Savva, Dening, & Brayne, 2012) and the chances of an individual developing one or more BPSD over the course of the dementia can be as high as 80% (Tible, Riese, Savaskan, & von Gunten, 2017). Many biological, psychological, and environmental factors contribute to the development and severity of BPSD (Freeman & Joska, 2012), which may be one reason there is no clear picture of the relationship between BPSD, the stages, and types of dementia. Considerable interindividual variability may contribute to the lack of a clear trajectory of BPSD as dementia severity increases (Milano et al., 2013). Ford (2014) cites research supporting the decrease of depressive and anxiety symptoms as dementia progresses and an increase in apathy, agitation, and irritability. However, a factor analytic approach by Potetti, Nuti, Cipriani, and Bonuccelli (2013) found low negative correlations between BPSD and level of cognitive impairment. Symptoms may also be related to the type of dementia that is diagnosed, further complicating treatment recommendations. Frequently, BPSD cause distress both to the individuals themselves and to their caregivers and contribute to a diminished quality of life. These symptoms are also major contributors to institutionalization (Milano et al., 2013; Kar, 2009).

## Music and Music-Therapy Interventions

The umbrella review described in this chapter defines a *music intervention* as the use of a musical experience by a practitioner or caregiver, such as listening, singing, playing rhythm instruments, moving to music, or discussion (i.e., music-based reminiscence) to address individual needs or to obtain a designated outcome. Music interventions may be passive/receptive, defined as listening to recorded or live music and/or active in which individuals are encouraged to participate in music activities. *Music therapy,* as defined by The American Music Therapy Association (AMTA), is "the clinical and evidence-based use of music interventions to accomplish individualized goals within a therapeutic relationship by a credentialed professional who has completed an approved music therapy program" (AMTA, n.d.). The primary difference between these definitions is the role of a credentialed professional in planning and implementing the music intervention.

## Methods

The recommended procedure for conducting an umbrella review involves "a clearly articulated question(s), detailed inclusion criteria, a structured

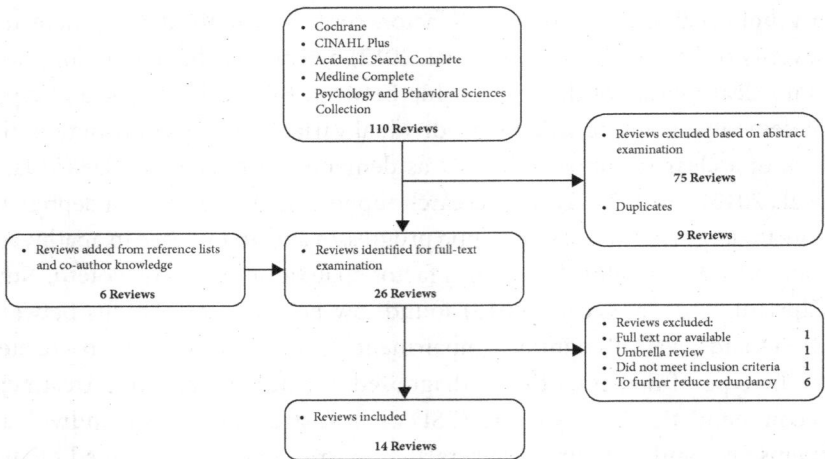

**Figure 10.1.** Review screening process.

search process to locate and select relevant existing reviews, method for critical appraisal of the included reviews and a formal process of data extraction
followed by means to summarize and present the data" (Aromataris et al.,
2015, p. 134). These recommendations were used to guide the umbrella review presented in this chapter.

## Search Strategy

Five databases were included in the search process, as shown in Figure 10.1.
Reference lists of identified reviews and the coauthors' knowledge of the professional literature also were used to identify possible reviews. Coauthors
searched each database separately, examined titles and abstracts, and
selected potential reviews for inclusion. Full texts of these selected reviews
were obtained for further evaluation. Search discrepancies were discussed
until mutual agreement was reached on which reviews would be included.
The following sets of terms were used to search the literature: *music or "music
therapy"* in combination with *dementia;* and *music or "music therapy"* in
combination with *"behavioral and psychological symptoms of dementia."* The
term *systematic review* also was included.

Systematic reviews were considered for inclusion if they examined the effects of music and/or music-therapy interventions (as defined in the Introduction) on BPSD, were written in English and published between 2000 and 2017 in peer-reviewed sources. There was no restriction on type of systematic review, and those which included music or music therapy as part of a review of general nonpharmacological interventions were considered if primary studies contained therein provided specific information about the type of music interventions used such that meaningful conclusions about their effectiveness could be drawn.

## Critical Appraisal

A priori, it was decided to use The Assessment of Multiple Systematic Reviews (AMSTAR) checklist (Shea et al., 2009, 2007) to evaluate the methodological quality of included reviews. An 11-item checklist, the AMSTAR is a reliable and valid tool specifically designed to assess the methodological integrity of systematic reviews (Pieper, Buechter, Li, Prediger, & Eikermann, 2015; Smith, Devane, Begley, & Clarke, 2011). Each item scoring a "yes" can be awarded 1 point, and a total score may be computed. While some umbrella reviews have translated AMSTAR scores into categorical ratings (Seo & Kim, 2012; van der Linde et al., 2012), no guidance for doing so is provided in the original research, and individual reviewers have handled this inconsistently (Burda, Holmer, & Norris, 2016). Given these limitations, a total score is provided for informational purposes only, but given the diversity in review characteristics, they were not categorically rated based on these scores.

## Data Extraction

Information extracted from each review included authors, year, and country of publication, professional credentials of authors, databases searched by review authors, number and design of primary studies, number and description of participants (including type and stage of dementia), BPSD studied, measurement tools for BPSD outcomes, and type of music intervention and interventionist(s). Data extraction was conducted independently by each co-author and discrepancies were resolved by discussion.

## Results

Results of the search process are outlined in Figure 10.1. The initial search yielded 104 reviews. Of these, 75 were initially excluded for the following reasons: not directly related to the topic (did not involve people with BPSD or did not include music as an intervention) or the focus of the review was on interventions for caregivers. Nine duplicates also were removed, and 6 reviews identified from reference lists and co-authors' knowledge of the literature were added, yielding a total of 26 possible reviews. Since the full text of one review was not available, copies of 25 reviews were obtained for full-text assessment. Following this assessment and further discussion, two reviews dealing with general nonpharmacological interventions for people with dementia were excluded because they did not specifically address music interventions for BPSD. Two reviews dealing with the use of music to enhance mealtime experiences among individuals in care facilities also were excluded because both reviews referred to the use of prerecorded sound or music as a mealtime adaptation (similar to the changing of seating patterns, modification of lighting, etc.). The focus of the music in these reviews was to modify the environment rather than to shape individual responses. To further reduce redundancy, a review was removed which appeared to be a condensed version of a larger review meeting inclusion criteria. Five additional reviews which contained less than 10 primary studies, all of which were discussed in larger reviews, also were excluded. Additionally, an umbrella review which contained descriptions of 38 systematic reviews addressing general nonpharmacological interventions for BPSD also was excluded. Since systematic reviews rather than individual studies were the primary data source in this review, it was decided that including it would have introduced unnecessary confusion.

Characteristics of included reviews are described in Table 10.1. The 14 reviews included in this overview were published between 2005 and 2017. The number of primary studies in each review ranged from 10 to 162 and were published between 1992 and 2015. Of these, 200 reported outcomes related to BPSD, but after accounting for primary studies included in multiple reviews, the revised total was 76 studies. Eight reviews used established criteria to evaluate the methodological quality of primary studies. Four reviews identified specific criteria for a diagnosis of dementia as part of the inclusion criteria (Chang et al., 2015; Ueda, Suzukamo, Sato, & Izumi, 2013; van der Steen et al., 2017; Vasionytė & Madison, 2013), and three reviews

**Table 10.1.** Characteristics of Included Reviews

| Author/Year/ Country and Author Credentials | Databases Searched by Original Review Authors | Primary Study Designs | Population | BPSD Symptoms | Measurement Tools for BPSD Outcomes | Type of Music Intervention(s) and Interventionist(s) | AMSTAR Rating |
|---|---|---|---|---|---|---|---|
| Chang et al. (2015) [Taiwan] Nurse (7), Physician (1), Gerontologist (1) | PubMed, Medline, Cochrane Library, CINAHL, Scopus, PsychINFO | N = 10; all RCTs | N = 465; Mild– severe stages of dementia (MMSE, GDS, CDR) Type: AD, VD, mixed) | Depression, Anxiety, Disruptive behaviors | CMAI, CMAI-SF, GDS, HARS, NPI, RAID, CSDD | Individual MT (preferred, recorded music), group MT (using musical instruments, recorded music) Interventionists: MT (5), Nurse (1), trained assistant (2), musician (1), NS (1) | 7 |
| Chatterton, Baker & Morgan (2010) [Australia] MT (2); NS (1) | Ebsco Host, Health Source (Nursing/ Academic Edition), Medline, Proquest, PubMed and 12 additional databases | N = 17; N = 4 dealt with "challenging behaviors" RCT (1); NCT (2); RCR (1) | N = 33; stages of dementia NS | Wandering, Agitation, Perseverative striking | Frequency counts using checklists; timing of duration of wandering | Music activities and singing Interventionists: MT (4) | 2 |

*(continued)*

Table 10.1. Continued

| Author/Year/ Country and Author Credentials | Databases Searched by Original Review Authors | Primary Study Designs | Population | BPSD Symptoms | Measurement Tools for BPSD Outcomes | Type of Music Intervention(s) and Interventionist(s) | AMSTAR Rating |
|---|---|---|---|---|---|---|---|
| Eells, K. (2014) [UK] Nurse | CINAHL, Medline, PsycARTICLES | N = 16; N = 6 studies on BPSD in dementia— RCT (1), RXO (1), QE (3), ABA (1) | N = 203; mild, moderate and severe dementia | Anxiety Agitation Depression Problem Behaviors | RAID; CMAI but not consistently reported | "Music therapy;" live group music; music listening; group singing. Interventionists: NS | 2 |
| Livingston et al. (2014) [UK] Author credentials NS | Medline, EMBASE, British Nursing Index, HTAP, PsychINFO, NHS Evidence, SIGL, SOOD, SNTIS, CINAHL, Cochrane Library | N = 160; N = 10 studies w/ music or MT protocol RCT (6); WS (2); NRCMC (1); NRXO (1); N = 11 studies w/ o music or MT protocol RC (5); WS (6); N = 13 studies no design indicated | N = 403 in studies w/ protocol; N = 278 in studies w/ o protocol; N = 296 in studies not meeting quality criteria; stage of dementia NS | Agitation | Outcome measures used in included studies not indicated | Group MT led by a trained therapist using a protocol which included a variety of music tasks. No protocol studies: listening to recorded music; Interventionists: NS | 9 |

| | | | | | | | |
|---|---|---|---|---|---|---|---|
| Livingston et al. (2005) [UK] Physician (3), NS (2) | Cochrane Library, "Electronic databases" NS | N = 162; N = 17 used music or MT to manage BPSDs RCT (4); Other designs NS | N = 413 in music + BPSD studies; stages of dementia NS | Depression, Agitation, Aggression, Disruptive vocalizations, Disruptive behavior, Irritability | Outcome measures used in included studies not indicated | Group MT, music (classical, stimulating/sedative, preferred), individual MT Interventionists: NS | 5 |
| McDermott et al. (2012) [UK] MT (1); NS (3) | Medline, Embase, PsychINFO, Cochrane Library, Web of Science | N = 18; N = 13 dealt with BPSD RCT (6); NRC (2); B&A (3); MM (2) | N = 411; moderate to severe dementia | BPSD (general); Depression; Anxiety; Aggression; Wandering; Agitation; | NPI, HARS, GDS, BEHAVE-AD, CMAI, STAI, CSDD, MOSES, DBRS | Improvisation-based group MT; individual receptive MT; singing; music listening; group MT (NS) Interventionists: MT (13) | 5 |
| OConnor et al. (2009) [Australia] Author credentials NS | Medline, CINAHL, PsychINFO, Cochrane | N = 25; N = 8 used music or MT; RCT (2); RXT (2); (4) RM = (2) | N = 226 in studies related to music; stages of dementia NS | Wandering, Aggression, Agitation, Verbal disruptions, Restlessness; "Challenging behaviors" | Measurement by direct observation and GBS | "Music therapy" (NS); Music listening (background, preferred classical relaxation music, baroque); natural sounds. Interventionists: NS | 6 |

(continued)

Table 10.1. Continued

| Author/Year/ Country and Author Credentials | Databases Searched by Original Review Authors | Primary Study Designs | Population | BPSD Symptoms | Measurement Tools for BPSD Outcomes | Type of Music Intervention(s) and Interventionist(s) | AMSTAR Rating |
|---|---|---|---|---|---|---|---|
| Petrovsky et al. (2015) [USA] Nurse (3) | Medline, PubMed, PsychINFO, CINAHL | N = 10; all dealt with BPSDs; RCT (3); Pre-post (6); RXO (1) | N = 391; Primarily mild, but mixed stages allowed | Anxiety Depression Mood | C-CSDD; CSDD; GDS; NPI-Q; RAID; MOSES; RMBPC Depression | Active music interventions including singing, playing instruments and drumming, movement. Interventionists: NS | 4 |
| Raglio et al. (2012) [Italy] Author credentials NS | PubMed, PsychINFO, Cochrane RCT Register | N = 32 studies; N = 20 dealt with BPSD. All were either RCT or CCT | N = 855 in studies dealing with BPSDs; mild to severe dementia | Depression; Anxiety; Agitation; Disruptive behavioral disturbances (NS); Apathy | CMAI; C-CMAI; GDS; HARS; BEHAVE-AD; CSDD, RAID | Active group music and/or MT (NS); music listening Interventionists: MT (8); OT (1); NS (11) | 3 |

| Author/Country | Databases | N studies | Population | Outcomes | Outcome tools | Interventions | |
|---|---|---|---|---|---|---|---|
| Ueda et al. (2013) [Japan] Author credentials NS | CINAHL, Medline, PsychINFO, Igaku Chuo Zasshi (Japanese only) | N = 20, N = 16 studies on BPSDs; RCT (8); CCT (8) | N = 540 in BPSD studies; stage of dementia: mild to severe; types: AD, vascular, Parkinson's type; MI | Anxiety Depression "Behavior & Neuropsychology" (p. 634) | RAID, HARS EDS, GDS, DS, CMAI, NPI, NPI-Q, JSS-D, JSS-E, BEHAVE-AD, GBS (Not clear what some outcome tools were) | Singing (familiar music), making musical instruments, hand bells, song drawing and writing, improvisation-based MT group, music and reminiscence, music listening (preferred, slow tempo and beat, Gerdner protocol, instrumental), individual MT (rhythm activity using preferred/familiar music), Clair & Bernstein Protocol, music exercise Interventionists: MT (9); nurse or nurse research assistant (4), clinical psychologist (1), caregiver (1), OT (1) | 8 |
| van der Steen et al. (2017) [The Netherlands] Author credentials NS | ALOIS (Specialized Cochrane database); Medinc; Embase; PsychINFO, CINAHL; Web of Science, LILACS, UMIN-CTRC, CENTRAL; Clinical Trials.gov; ICTRP Search Portal | N = 17 RCTs; N = 13 dealt with BPSDs | N = 642 in studies dealing with BPSDs; mild to severe dementia (DSM IV, DSM5 or ICD10 criteria; physician's dx or other tool acceptable | BPSD (general), Agitation, Anxiety, Aggression, Depression | CMAI, C-CMAI, C-CSDD, GDS, NPI, BEHAVE-AD, RAID | Active group music therapy and general music interventions; music listening Interventionists: MT (9), NS (4) | 10 |

*(continued)*

Table 10.1. Continued

| Author/Year/ Country and Author Credentials | Databases Searched by Original Review Authors | Primary Study Designs | Population | BPSD Symptoms | Measurement Tools for BPSD Outcomes | Type of Music Intervention(s) and Interventionist(s) | AMSTAR Rating |
|---|---|---|---|---|---|---|---|
| Vasionytė & Madison (2013) [Sweden] Author credentials NS | JSTOR, EBSCO, ERIC, SCIRUS, Medline, PsychINFO, Cochrane Library, ProQuest | N = 19; N = 12 dealt with BPSDs; RCT (4); CCT (4); RXO (1); QE (3) | N = 304 in studies dealing with BPSDs; mild to severe dementia; AD, vascular, general (clinical dx. or established diagnostic) | "Behavioral" and "Affective," BPSDs NS | CMAI, NPI-Q, RAID, GDS, MOSES | "Active" and "passive," live/ recorded music; group/ individual approaches Interventionists: NS | 6 |
| Wall & Duffy (2010) [Ireland, UK] Nurse (2) | CINAHL, PsychINFO, Medline | N = 13; N = 11 dealt with BPSDs; RCT (3); CT (3); RXO (1); TRS (1); NS (1) | N = 268; stage of dementia not consistently specified | Agitation; Behavioral problems | CMAI; NPI; BEHAVE-AD | Group MT sessions; recorded background music; music + movement; preferred recorded music Interventionists: MT (5), Nurse and trained assistant (1), and NS (5) | 3 |

| | | | | | |
|---|---|---|---|---|---|
| Zhang et al. (2017) [PR China] Author credentials NS | PubMed, Embase, Cochrane Library | N = 34; N = 26 dealt with BPSDs; RCT (14); CCT (7); RXO (5) | N = 1329; All stages of dementia included | Disruptive behavior, Depression, Anxiety | CMAI; BEHAVE-AD; NPI; NPI-Q; GDS, CSDD; RAID; | Interactive and passive music intervention. Interventionists: NS |

8

Key: Databases Searched: ALOIS = Specialized Cochrane database; CINAHL = Cumulative Index of Nursing and Allied Health Literature; EBSCO Host = eBook Academic Collection; ERIC = Education Resources Information Center; HTAP = Health Technology Assessment Programme; ICTRP = International Clinical Trials Registry Platform; JSTOR = Journal Storage; LILACS = Latin American and Caribbean Health Sciences; SIIT = Scirus (comprehensive science-specific search engine); SIGL = System for Information on Grey Literature; SOOD = Stationery Office Official Documents website; SNTIS = Stationery National Technical Information Service; UMIN-CTRC = UMIN Clinical Trials Registry

Key: Primary Study Designs: ABA = Applied Behavior Analysis; B&A = Before and after; CCT = clinical controlled trial; CT = clinical trial; MM = mixed methods; NCT = nonrandomized controlled trial; NRC = nonrandomized controlled studies; NRCMC = nonrandomized case match control study; NRXO = nonrandomized cross-over trial; QE = quasi-experimental; RCR = retrospective clinical report; RCT = randomized controlled trial; RM = repeated measures; RXO = randomized crossover trial; RXT = repeated measures w/randomized crossover; TRS = Triangulated Research Study; WS = within subjects

Key: Measurement Tools for levels of dementia: MMSE = Mini-Mental State Exam; GDS = Global Dementia Scale; CDR = Clinical Dementia Rating

Key: Measurement Tools for BPSD: BEHAVE-AD = Behavioral Pathology in Alzheimer's Disease Rating Scale; CMAI = Cohen-Mansfield Agitation Inventory; C-CMAI = Chinese version of the CMAI; CMAI-SF = CMAI—Short Form; CSDD = Cornell Scale for Depression in Dementia; C-CSDD = Chinese version of the CSDD; DS = Depression Scale; DBRS = Disruptive Behavior Rating Scales; EDS = Emotional Disturbance Scale; GBS = Gottfries-Bråne-Steen scale; GDS = Geriatric Depression Scale; HARS = Hamilton Anxiety Rating Scale; JSS-D = Japan Stroke Scale—Depression; JSS-E = Japan Stroke Scale = Emotional Disturbance; MOSES = Multidimensional Observation Scale for Elderly Subjects; NPI = Neuropsychiatric Inventory; NPI-Q = Neuropsychiatric Inventory—Brief Questionnaire Form; RAID = Rating Anxiety in Dementia Scale; RMBPC = Revised Memory and Behavioral Problems Checklist

Key: Type of music intervention(s) and Interventionist(s): MT = music therapy/music therapist; NS = not specified; OT = occupational therapist included information on type of dementia diagnosed (Chang et al., 2015; Ueda et al., 2013; Vasionytė & Madison, 2013). A variety of outcome measures were used in primary studies and two reviews (Chang et al., 2015 and Ueda et al., 2013) specified that outcome measures selected for their review had to have psychometrically supported reliability and validity. The AMSTAR ratings ranged from 2 to 10 which further illustrates the diversity in purpose and focus of each review.

## Effects of Music or Music-Therapy Interventions on BPSD

The next section presents findings from the included reviews on the effectiveness of music and/or music-therapy interventions in addressing BPSD. Findings are organized by symptom and following a definition of the symptom, conclusions from relevant reviews are summarized. In reviews that included a meta-analysis, statistical tests also were performed to examine the *heterogeneity* or degree of variability between individual studies included in the analysis. Sources of heterogeneity and their impact on the interpretation of review results are described in the Discussion section. Detailed statistical information on results of meta-analyses can be found by consulting the original reviews.

Ten systematic reviews reported on the effects of music or music-therapy intervention on overall BPSD. Five of these reviews included a meta-analysis (Chang et al., 2015; Ueda et al., 2013; van der Steen et al., 2017; Vasionytė & Madison, 2013; Zhang et al., 2017). In order to conduct these analyses, individual studies examining the effects of music or music-therapy interventions on different BPSD were combined into a primary analysis to produce one effect size. Primary studies used tools such as the Neuropsychiatric Inventory (NPI) or the Behavioral Pathology in Alzheimer's Disease Rating Scale (BEHAVE-AD) to report outcomes; details on measurement tools used by these authors are presented in Table 10.1. When a review included a sufficient number of studies on individual BPSD, additional analyses were performed; results of these additional analyses are reported in the next three sections.

Primary analyses in two reviews (Chang et al., 2015; Zhang et al., 2017) found significant effects in favor of music or music-therapy interventions in decreasing BPSD. Additional regression analysis conducted by Zhang et al. (2017) identified intervention method, comparator, and trial design as possible sources for heterogeneity. Subgroup analysis conducted by Chang et al. (2015) showed stronger effects for group music interventions and for individual music sessions offered twice a week for participants with mild to moderate dementia. A review by Ueda and colleagues (2013) found a small effect for overall music intervention with some study heterogeneity. Subgroup analyses showed moderate effects for music listening and singing. Van der Steen et al. (2017) in their review found no effects for therapeutic music intervention at end of treatment or for up to 1-month follow-up. Heterogeneity was low to moderate for end-of-treatment effects and not important for effects at follow-up. The effect size for music or music-therapy intervention on behavioral outcomes reported by Vasionytė and Madison (2013) was interpreted as large but nonsignificant. The behavioral outcomes reported in this review were measured by a variety of instruments as indicated in Table 10.1.

Five reviews presented findings related to the effects of music or music-therapy interventions on BPSD through narrative description. McDermott, Crellin, Ridder, and Orrell (2013) found evidence in support of music therapy for short-term improvement in BPSD. Raglio et al. (2012) reviewed 20 primary studies dealing with BPSD and concluded that most music-therapy interventions, primarily those involving active participation, led to an improvement in BPSD when compared with usual care or no treatment. Chatterton, Baker and Morgan (2010) reported evidence from four studies demonstrating that singing by either practitioners or caregivers was effective in decreasing challenging behaviors such as wandering, perseverative striking, agitation, and isolating behavior among people with dementia. O'Connor, Ames, Gardner, and King (2009) summarized eight studies that examined the effectiveness of music or sound to decrease a variety of BPSD and concluded that music, especially live and preferred music, was very effective in five out of the eight studies evaluated. The review of nonpharmacological interventions for BPSD by Livingston et al. (2005) is the earliest in the present overview and included 24 primary studies using music interventions, of which 17 specifically examined a variety of BPSD outcomes. Their evaluation of these studies concluded that there is some evidence for the effectiveness of music or music therapy in reducing problem behaviors, especially agitation, during and immediately after sessions. In summary, the evidence suggests that music and music-therapy interventions, particularly those that are live and are offered in a group context, are successful for short-term reduction of BPSD.

## Effects of Music or Music-Therapy Interventions on Agitation

Agitation is one of the most common behavioral symptoms exhibited by people with dementia (Kong, Evans, & Guevara, 2009). It is defined as "an inappropriate verbal, vocal, or motor activity that is not explained by apparent needs or confusion" (Van der Mussele et al., 2015, pg. 247). Symptoms of agitation can include abuse or aggressive behavior toward self or others, appropriate behaviors performed with inappropriate frequency, or behaviors that are inappropriate according to social standards. Other symptoms such as anxiety, panic, depression, delusions, hallucinations, and/or delirium may accompany agitation (Lou, 2001). The causes of agitation can be related to neurological, physical, psychological, functional, interpersonal relational, environmental, and restraint factors (Kong, Evans, & Guevara, 2009). Furthermore, clinicians have been able to identify three subtypes of agitation: (a) physically nonaggressive behavior, such as wandering or trespassing in inappropriate places; (b) aggressive behavior, such as hitting and kicking; and (c) and verbally agitated behavior such as repeating words or questions, demanding constant attention, shouting, or verbal aggression (Millán-Calenti et al., 2016; Livingston et al., 2014). Symptoms of agitation may be present in more than 80% of people with dementia who reside in residential care homes, and agitation can be detrimental to an individual's safety, security, and quality of life (Vink, Zuidersma, Jonge, Zuidema, & Slaets, 2012).

Three reviews examined the effects of music or music-therapy interventions on agitation. Wall and Duffy (2010) concluded that music therapy reduced agitation levels and promoted a positive increase in mood and socialization among people with dementia, the effects of which also extended to caregivers. They further concluded that the use of live, individualized music was most beneficial in minimizing verbally aggressive and non-physically aggressive behavior (Wall & Duffy, 2010). Livingston et al. (2014) found that music therapy by protocol was effective in decreasing emergent agitation and decreasing symptomatic agitation but showed no long-term benefits and no evidence of effectiveness for people with severe agitation. Livingston et al. (2014) described a *protocol* as a music intervention which included a well-known warm-up song followed by a structured series of receptive and/or active music experiences (p. 20). Van der Steen el al. (2017) found no clear evidence for the effectiveness of music or music therapy in decreasing either agitation or aggression. In summary, the evidence for the effects of music and music-therapy intervention on agitation is mixed. These mixed results are likely due to different methodologies used by reviewers in selecting and rating primary studies and to sources of heterogeneity within those studies. The issue of heterogeneity is further addressed in the Discussion section below.

## Effects of Music or Music-Therapy Interventions on Anxiety

Anxiety may be defined as "apprehensive fear and excessive worry" (Eells, 2014, p. 10). Symptoms of anxiety may include worry, restlessness, irritability, muscle tension, shortness of breath, and difficulty in concentrating, relaxing, or sleeping (Neville & Teri, 2011; Seignourel, Kunik, Wilson, & Stanley, 2008). Anxiety symptoms are relatively consistent across all levels of dementia severity until the terminal stage where symptoms decrease. Symptoms of anxiety are most noted in vascular dementia, frontotemporal dementia, and Parkinson's-related dementia when compared with Alzheimer's disease (Freeman & Joska, 2012). For people living with dementia, anxiety may interfere with basic activities of daily living, may be accompanied by other problem behaviors, such as wandering and verbal threats, and may lead to increased residential care home or assisted living placement (Petrovsky et al., 2015).

Five reviews, two of which included meta-analyses, reported on the effects of music or music-therapy interventions on anxiety for people with dementia. Results of a meta-analysis by Chang et al. (2015) found that music therapy lowered anxiety levels, decreased disruptive behaviors and improved mood for people with dementia. This review also concluded that group interventions offered twice a week for people with mild to moderate dementia were most effective in reducing anxiety. In a review and meta-analysis by Zhang et al. (2017), interactive music interventions had short-term positive effects on anxiety. Eells (2014) indicated that music is beneficial in reducing anxiety among older adults with dementia and their caregivers and that it can be used "easily and effectively as a therapeutic nursing intervention" (p. 17). However, Eells (2014) cautioned that small sample sizes, differences in length and frequency of music-therapy sessions, and varied utilization of music interventions make it difficult to substantiate this conclusion. In the review by Petrovsky et al. (2015), results were inconclusive regarding the effects of music interventions on anxiety in people with mild dementia. The lack of sufficient evidence to support the effectiveness of music was most likely due to low methodological rigor of primary studies, but this was not formally assessed. A review by van der Steen et al. (2017) concluded that anxiety was lower in the music intervention groups at the end of treatment, but the evidence in support of this outcome was deemed to be of low quality. In summary, the evidence suggests that music and music-therapy interventions may be effective in reducing symptoms of anxiety among people in mild or moderate stages of dementia. Inconclusive results may be due to differences in methodology and in quality ratings of primary studies by review authors.

## Effects of Music or Music-Therapy Interventions on Depression

Six reviews reported on the effects of music or music-therapy interventions in addressing depression in dementia. Depression is characterized by a pervasive sad or low mood, a decrease in energy, feelings of hopelessness or pessimism and physical changes such as loss of appetite or sleep difficulties (Ganguli, 2009). Since cognitive changes may be symptomatic of depression and a history of depression may be a risk factor for the development of dementia, a thorough evaluation of depressive symptoms is imperative if dementia is suspected (Muliyala & Varghese, 2010).

Two meta-analyses found moderate effects for music or music-therapy interventions on end of treatment depression outcomes among people with dementia (Chang et al., 2015; van der Steen et al., 2017), while effects at 1- to 4-month follow-up measurements were low (van der Steen et al., 2017). Zhang et al. (2017) found no significant differences between music or music-therapy interventions and controls in decreasing depression but trends favored music therapy, with similar results for both interactive and passive interventions. A meta-analysis by Ueda et al. (2013) found a small effect in favor of music or music-therapy intervention for depressive symptoms among people with dementia; subgroup analyses showed a moderate effect for singing interventions. A medium but nonsignificant effect was found for music or music-therapy interventions on affect among participants with dementia in a meta-analysis by Vasionytė and Madison (2013). Six studies were included in this category, three of which dealt specifically with depression. After presenting narrative descriptions of nine studies of music or music-therapy interventions on symptoms of depression among participants with mild dementia, Petrovsky et al. (2015) concluded that there was insufficient evidence to support the effectiveness of music interventions in decreasing depression among these individuals. In summary, the evidence indicates that both interactive and passive music and music-therapy interventions may be effective for short-term reduction in symptoms of depression. Inconclusive results may be due to differences in methodology and in quality ratings of primary studies by review authors.

## Discussion

The purpose of this umbrella review was to summarize and compare the evidence in existing systematic reviews on the effectiveness of music and music-therapy interventions in addressing BPSD. Findings from 9 of 10 reviews concluded that music and music-therapy interventions had a positive effect on short-term reduction of BPSD. However, when effects of music and music-therapy interventions on anxiety, agitation, and depression were examined separately, results were mixed. This difference is likely due to the smaller number of primary studies included in separate analyses as well as the sources of diversity among the reviews and among the primary studies selected for inclusion. Sources of heterogeneity among reviews and their possible effects on findings are further discussed following the recommendations.

## Recommendations for the Use of Music and Music-Therapy Interventions on BPSD

Based on the conclusions of the reviews in this overview, recommendations for clinical best practice for the use of music and music-therapy interventions to address BPSD among people with dementia include:

1. the use of individualized, preferred music (O'Connor et al., 2009; Raglio et al., 2012; Wall & Duffy, 2010);
2. the use of live music (O'Connor et al., 2009; Wall & Duffy, 2010);
3. group music therapy offered several times a week for those with mild to moderate dementia (Chang et al., 2015); and
4. the use of active, protocol-based music interventions or music therapy (Livingston et al., 2014; Raglio et al., 2012; Zhang et al., 2017).

Increasingly, a person-centered approach is recognized as best practice in the care of people with dementia. The Institute of Medical Core Competencies defines person-centered care as "care that is respectful and responsive to individual patient preferences, needs, and values, and ensuring that patient values guide all clinical decisions" (Morgan & Yoder, 2012, p. 8). In this approach, BPSD are understood not as disturbances requiring medication, but as manifestations of unmet psychological needs or as attempts to communicate these needs (Ridder & Wheeler, 2015). When music and music-therapy interventions are provided within the context of a person-centered approach, there is an increased likelihood that BPSD will be reduced.

## Sources of Heterogeneity

**Study quality.** Several reviews limited study design to RCTs or CCTs (Chang et al., 2015; Raglio et al., 2012; Ueda et al., 2013; van der Steen et al., 2017), while other reviews did not (McDermott et al., 2013; Wall & Duffy, 2010). Study quality was rated in eight reviews and a variety of criteria were used for this assessment. Some reviews used study quality as a basis for inclusion in meta-analyses (Chang et al., 2015; Ueda et al., 2013; Zhang et al., 2017) but a discussion of study quality was not integrated into these reviews' conclusions. In some reviews, quality ratings were provided in tables of study characteristics, but no explanation for categorical descriptions such as "good," "fair," or "high quality" was provided (Chang et al., 2015; McDermott et al., 2013). Some reviews provided concise descriptions of the methods selected to rate study quality and integrated study quality into conclusions (Livingston et al., 2014; Livingston et al., 2005; McDermott et al., 2013; van der Steen et al., 2017). One review (O'Connor et al., 2009) provided a clear description of the checklist used to rate study quality which included information related to how categorical ratings were assigned, but observations on quality ratings were not integrated into the conclusions. Both the Forbes checklist (Forbes, 1998 cited in O'Connor et al., 2009) and the Downs and Black checklist (Downs & Black 1998 cited in McDermott et al., 2013) are sensitive to study designs other than RCTs and appear to provide more flexibility for rating studies on the effects of music interventions than do criteria that rate study quality using a strict hierarchy of evidence (see Livingston et al., 2014). The PEDro scale was selected to rate study quality in the review by Zhang et al. (2017). PEDro is the Physiotherapy Evidence Database, a free service designed to provide practitioners with quality evidence to support best practice (pedro.org.au, n.d.). Its 11-item checklist is frequently used to evaluate internal validity of primary studies included on its database (pedro. org.au/wp-content/uploads/PEDro_scale.pdf). A rationale for the choice of this tool by Zhang et al. (2017) was not provided. Diversity in the approaches used by review authors to evaluate study quality presents an additional challenge to formulating clear conclusions about the effectiveness of music and music-therapy interventions in BPSD. Information on study quality appears to be most useful when methods used to obtain ratings are clearly described and integrated into a review's conclusions.

**Study participants.** As seen in Table 10.1, there was considerable diversity in both type and stage of dementia among participants in primary studies included in each review. In reviews where support for the effectiveness of music interventions was found, these effects lasted for the duration and immediately following the intervention. Given the nature of dementia as a neurodegenerative disorder, the failure to find evidence for longer-term effects is not surprising. Subgroup analysis by Chang et al. (2015) found that music therapy had a larger and more positive effect on behavioral disturbances and depression for people with mild to moderate dementia than for those with moderate to severe dementia. They also found that effect sizes for music therapy were larger for anxiety in those with moderate to severe dementia than for those with mild to moderate dementia. Ueda et al. (2013) observed that because many studies are conducted in group settings, controlling for severity of dementia may be difficult. While three reviews (Chang et al., 2015; Ueda et al., 2013; Vasionytė & Madison, 2013) included information on types of dementia diagnosed in primary study participants, the effects of music or music-therapy intervention on type of dementia was not examined in any of the analyses.

**Outcome measurement tools.** Differences in outcome measurement tools was another source of heterogeneity. Petrovsky et al. (2015) noted that the variety of outcome measures used in primary studies may complicate the ability to interpret results. They further reported that for people with mild dementia, the Geriatric Depression Scale (GDS) is considered appropriate for measuring depression, but that scales measuring anxiety vary widely, and there is little agreement about which may be the most appropriate tool for measuring anxiety. Two reviews (Chang et al., 2015; Ueda et al., 2013) specified that outcome measures in primary studies selected for their review had to have psychometrically supported reliability and validity. Robustness of reliability and validity measures was used to evaluate study quality in two reviews (Livingston et al., 2014; McDermott et al., 2013). Zhang et al. (2017) reported that differences in measurement tools contributed to substantial heterogeneity in subgroup analysis and meta-regression for effects of music intervention on disruptive behaviors. McDermott et al. (2013) noted that in order to improve trustworthiness of results, researchers may select well-validated measures, yet some of these may not be sensitive to music-therapy treatment effects. For example, tools such as the Behavioral Pathology in Alzheimer's Disease Rating Scale (BEHAVE-AD) and the Neuropsychiatric Inventory (NPI) are global measures which assess broad categories of symptoms. These two measures differ in their approaches to assessing BPSD and have been used primarily in clinical trials to evaluate the effectiveness of pharmacological interventions (Reisberg et al., 2014). The Behavioral Pathology in Alzheimer's Disease Rating Scale (BEHAVE-AD) has the advantage of assessing the severity and frequency of individual symptoms whereas the Neuropsychiatric Inventory (NPI) assesses the severity and frequency of *categories* of symptoms (Reisberg et al., 2014). Comparisons by Ing-Randolph, Phillips, and Williams (2015) further illustrate how choice of different outcome measures may contribute to study heterogeneity. They recommend use of the Rating Anxiety in Dementia Scale (RAID) to measure anxiety in those with mild to moderate dementia, primarily because it not only is well standardized, but includes a self-report component. McDermott et al. (2013) also emphasized the importance of using self-report measures as some people with dementia can provide this information; proxy and observational measures may limit the perspective of the person him or herself.

**Definitions of music therapy.** Another source of heterogeneity is the inconsistent use of the term "music therapy." Only one review had specific inclusion criteria related to music therapy (McDermott et al. 2013). They used a detailed definition of music therapy and the authors described the difficulty in differentiating "music activity" from "music therapy" in many of the primary studies they reviewed. Two reviews (Zhang et al. 2017; Wall & Duffy, 2010) cited the American Music Therapy Association's (AMTA's) definition of music therapy but the use of a credentialed professional to deliver the music was not specified in the primary studies they included, nor was it mentioned in their discussion as a possible factor influencing outcomes. In their conclusions, Chang et al. (2015) noted the importance of specialized music-therapy training and credentials, but also considered music an intervention that nurses can provide. A sensitivity analysis conducted by Van der Steen et al. (2017) found that the use of a credentialed music therapist to provide the music interventions did not significantly improve effect sizes, but because of the small number of primary studies in this analysis, the results may be inconclusive. Inconsistency in the definitions of music therapy seen in these reviews creates confusion regarding specialized credentials needed to provide different types of music intervention. Chatterton et al. (2010) offered some supporting evidence to help clarify these differences. For instance, they noted that music therapy is goal-oriented and tailored to meet individual needs. In contrast, professional caregivers or family members may use music interventions to build or maintain meaningful interpersonal connections. Both act within a person-centered model but use different approaches. Since this distinction is not evident in the reviews examined for this overview, conclusions regarding the effectiveness of interventions to address BPSDs which are delivered by a credentialed music therapist versus those delivered by others remain inconclusive. Clarity and consistency might be enhanced by including credentialed music therapists in research teams.

   **Music and music-therapy interventions.** The variety of music interventions featured in primary studies is another source of heterogeneity. In the review by Ueda et al. (2013), 15 primary studies used a combination of interventions such as singing, playing musical instruments, or listening to live performances. Vasionyté and Madison (2013) included studies that involved active or receptive music therapy and music listening. There were only two reviews (Chatterton et al., 2010; Eells, 2014) that focused on single interventions emphasizing singing. During music-therapy sessions, it is standard practice to offer a variety of engagement opportunities for people in different stages of dementia. However, there are instances in which sessions may be designed to meet specific needs, such as singing to increase alert responses in people with late-stage dementia (Clair, 1996), or singing familiar songs to people in end-stage dementia to provide comfort or reassurance. It is important for practitioners to consider the severity of dementia and to adjust goals according to individual need at any point in the therapeutic process (Cevasco & Grant, 2006). Due to the complex nature of the various types of dementia, much remains to be understood about the relationship between type and severity of dementia and the effects of music therapy (Chang et al., 2015). Future studies should examine in more detail the effects of music interventions and music therapy on specific BPSD on people in various stages of dementia.

## Conclusions

Almost 90% of people living with dementia experience BPSD (Raetz, 2013), which contribute to a diminished quality of life, both for the person affected and their caregivers, especially in the advanced stages of the disease (Raglio et al., 2015). The purpose of this umbrella review was to summarize and compare the evidence in existing systematic reviews on the role of music and music-therapy interventions in addressing BPSD among individuals living with dementia. The reviews included share a consensus that these interventions are a noninvasive, safe, and useful alternative to medications in reducing BPSD. Our review concurs with the results of an umbrella review of 38 systematic reviews of nonpharmacological interventions (of which music therapy was one intervention), which concluded that music therapy was a promising intervention for BPSD (Abraha et al., 2017). While Abraha et al. (2017) provided brief descriptive summaries of each review they included, the present overview focused only on the effectiveness of music and music-therapy interventions in addressing BPSD and has emphasized the implications of review results for practitioners desiring to use these interventions for their clients with dementia.

The diversity that characterizes the primary studies represented in the 14 reviews discussed in this overview has both strengths and limitations. A strength is that this diversity may provide a holistic perspective from which to examine the phenomenon. Methodological variations may be considered a limitation in that they complicate the ability to conduct meta-analyses and to use these analyses to formulate precise conclusions about effectiveness. However, O'Connor et al. (2009) emphasize the importance of balancing methodological rigor with clinical relevance. The issue of what constitutes acceptable evidence for the efficacy of music and music-therapy interventions has been addressed by Abrams (2010) and Aigen (2015). Abrams (2010) argues that a broader understanding of what is meant by "evidence" needs to include the clinical expertise of the practitioner as well as the lived experience of the client or patient. His model identifies four distinct epistemological perspectives from which evidence may be derived. These consist of *objective* (exterior-individual), *interobjective* (exterior-collective), *subjective* (interior-individual), and *intersubjective* (interior-collective) perspectives (Abrams, 2010, p. 353). Abrams contends that evidence for effective music-therapy practice needs to accommodate all of the perspectives represented in this model. In his critique of traditional definitions of evidence-based practice as applied to music therapy, Aigen (2015) argues that "the medical model and its associated illness ideology" may be too restrictive for a full understanding of what makes music therapy work (p. 12). Newer approaches such as culture-centered music therapy (Stige, 2002) and resource-oriented music therapy (Rolvsjord, 2010) are emerging and are challenging traditional, clinical definitions of the discipline. Aigen (2015) further argues that social contexts, the empowerment of clients, and other humanistic factors need to be considered in determining efficacy of any therapeutic approach. None of these factors were assessed in any of the reviews included in this overview.

Credentialed music therapists have the knowledge and training to assess, implement, and evaluate the effectiveness of music interventions to address BPSD. Their expertise can guide other professionals in the use of these interventions so that the comfort and well-being of people with dementia is optimized and potential harm avoided. A thorough assessment of musical preferences and experiences can determine which music interventions are most likely to meet the needs of a person with dementia and which might increase agitation or anxiety. Because music can elicit powerful emotions, practitioners or caregivers also need to have the skills to manage emotions that people with dementia may express during a music experience and to guide the person to a calm or safe space when necessary.

Given the considerable interindividual differences in the trajectory of BPSD associated with type or stage of dementia, general recommendations for the use of music or music-therapy interventions for specific symptoms at any one stage are difficult to make. While Clair and Memmott (2008) do offer suggestions for music interventions to manage specific behavioral difficulties over the course of dementia, best practice dictates that individualized music and music-therapy interventions delivered within the context of a person-centered approach are those likely to be the most successful.

# References

References marked with an asterisk indicate reviews included in overview.

Abraha, I., Rimland, J. M., Trotta, F. M., Dell'Aquila, G., Cruz-Jentoft, A., Petrovic, M., . . . Cherubini, A. (2017). Systematic review of systematic reviews of non-pharmacological interventions to treat behavioral disturbances in older patients with dementia. The SENATOR-On Top series. *BMJ Open, 7,* e012759. doi:10.1136/bmjopen-2016-012759

Abrams, B. (2010). Evidence-based music therapy practice: An integral understanding. *Journal of Music Therapy, 47(4),* 351–379.

Aigen, K. (2015). A critique of evidence-based practice in music therapy. *Music Therapy Perspectives, 33(1),* 12–24.

Aromataris, E., Fernandez R., Godfrey C. M., Holly, C., Khalil, H., & Tungpunkom, P. (2015). Summarizing systematic reviews: Methodological development, conduct and reporting of an umbrella review approach. *International Journal of Evidence-based Healthcare, 13(3),* 132–140.

Burda, B. U., Holmer, H. K., & Norris, S. L. (2016). Limitations of a measurement tool to assess systematic reviews (AMSTAR) and suggestions for improvement. *Systematic Reviews*, 5:58. doi:10.1186/s13643-016-0237-1

Cevasco, A. M., & Grant, R. E. (2006). The value of musical instruments used by the therapist to elicit responses from individuals in various stages of Alzheimer's disease. *Journal of Music Therapy*, 43(4), 226–246.

Cerejeira, J., Lagarto, L., & Mukaetova-Ladinska, E.B. (2012). Behavioral and psychological symptoms of dementia. *Frontiers in Neurology, 3, Article 73*, 1–21; doi:10.3389/fneur.2012.00073

*Chang, Y., Chu, H., Yang, C., Tsai, J., Chung, M., Liao, Y., . . . Chou, K. (2015). The efficacy of music therapy for people with dementia: A meta-analysis of randomised controlled trials. *Journal of Clinical Nursing, 24*, 3425–3440, doi:10.1111/jocn.12976

*Chatterton, W., Baker, F., & Morgan, K. (2010). The singer or the singing: Who sings individually to persons with dementia and what are the effects? *American Journal of Alzheimer's Disease & Other Dementias, 25(8)*, 641–649.

Clair, A. A., & Memmott, J. (2008). *Therapeutic uses of music with older adults* (2nd ed.). Silver Spring, MD: American Music Therapy Association.

Clair, A. A. (1996). The effect of singing on alert responses in persons with late stage dementia. *Journal of Music Therapy, 33(4)*, 234–247.

Downs, S. H., & Black, N. (1998). The feasibility of creating a checklist for the assessment of the methodological quality both of randomised and non-randomised studies of healthcare interventions. *Journal of Epidemiology Community Health, 52(6)*, 377–384.

*Eells, K. (2014, February). The use of music and singing to help manage anxiety in older adults. *Mental Health Practice, 17(5)*, 10–17.

Forbes, D. A. (1998). Strategies for managing behavioural symptomatology associated with dementia of the Alzheimer type: A systematic overview. *Canadian Journal of Nursing Research, 30*, 67–86.

Ford, A. H. (2014). Neuropsychiatric aspects of dementia. *Maturitas, 79*, 209–215. http://dx.doi.org/10.1016/j.maturitas.2014.04.005.

Freeman, C. P., & Joska, J. (2012). Management of behavioral and psychological symptoms of dementia. *Continuing Medical Education, 30(4)*, 110–113.

Ganguli, M. (2009). Depression, cognitive impairment and dementia: Why should clinicians care about the web of causation? *Indian Journal of Psychiatry, 51(Suppl 1)*, S29–S34, PMID: 21416013.

Ing-Randolph, A. R., Phillips, L. R., & Williams, A. B. (2015). Group music interventions for dementia-associated anxiety: A systematic review. *International Journal of Nursing Studies, 52*, 1775–1784.

Kar, N. (2009). Behavioral and psychological symptoms of dementia and their management. *Indian Journal of Psychiatry, 51*, 577–586.

Kong, E. H., Evans, L. K., & Guevara, J. P. (2009). Non-pharmacological intervention for agitation in dementia: A systematic review and meta-analysis. *Aging & Mental Health*, 13(4), 512–520.

*Livingston, G., Johnston, K., Katona, C., Paton, J., & Lyketsos, C. (2005). Systematic review of psychological approaches to the management of neuropsychiatric symptoms of dementia. *American Journal of Psychiatry, 162*, 1996–2021.

*Livingston, G., Kelly, L., Lewis-Holmes, E., Baio, G., Morris, S., Patel, N., . . . Cooper, C. (2014). A systematic review of the clinical effectiveness and cost-effectiveness of sensory, psychological and behavioural interventions for managing agitation in older adults with dementia. *Health Technology Assessment, 18(39)*, 1–256.

Lou, M. F. (2001). The use of music to decrease agitated behaviour of the demented elderly: The state of science. *Scandinavian Journal of Caring Science, 15*, 165–173.

*McDermott, O., Crellin, N., Ridder, H., & Orrell, M. (2013). Music therapy in dementia: A narrative synthesis systematic review. *International Journal of Geriatric Psychiatry, 28*, 781–794.

Milano, W., Saturnino, C., & Capasso, A. (2013). Behavioral and psychological symptoms of dementia: An overview. *Current Neurobiology 4*(1 & 2), 31–34.

Millán-Calenti, J.C., Lorenzo-López, L., Alonso-Búa, B., de Labra, C., González-Abraldes, I., & Maseda, A., (2016). Optimal nonpharmacological management of agitation in Alzheimer's disease: Challenges and solutions. *Clinical Interventions in Aging, 11*, 175–184.

Morgan, S., & Yoder, L. H. (2012). A concept analysis of person-centered care. *Journal of Holistic Nursing, 30*(1), 6–15.

Muliyala K., & Varghese, M. (2010). The complex relationship between depression and dementia. *Annals of Indian Neurology, 13*, S69–S73.

Neville, C., & Teri, L. (2011). Anxiety, anxiety symptoms, and associations among older people with dementia in assisted-living facilities. *International Journal of Mental Health Nursing, 20*, 195–201.

*O'Connor, D. W., Ames, D., Gardner, B., & King, M. (2009). Psychological treatments of behavioral symptoms in dementia: A systematic review of reports meeting quality standards. *International Psychogeriatrics, 21*(2), 225–240. doi:10.1017/S1041610208007588

*Petrovsky, D., Cacchione P. Z., & George, M. (2015). Review of the effect of music interventions on symptoms of anxiety and depression in older adults with mild dementia. *International Psychogeriatrics, 27*(10), 1661–1670. doi:10.1017/S1041610215000393

Pieper, D., Buechter, R. B., Li, L., Prediger, B., & Eikermann, M. (2015). Systematic review found AMSTAR, but not R(evised)-AMSTAR, to have good measurement properties. *Journal of Clinical Epidemiology, 68*, 574–583.

Poletti, M., Nuti, A., Cipriani, G., & Bonuccelli, U. (2013). Behavioral and psychological symptoms of dementia: Factor analysis and relationship with cognitive impairment. *European Neurology, 68*, 76–82. doi:10.1159/000341956

Raetz, J. (2013). A nondrug approach to dementia. *Journal of Family Practice, 62*(10), 548–557.

Raglio, A., Bellandi D., Baiardi P., Gianotti M., Ubezio M. C., Zanacchi E., . . . Stramba-Badiale, M. (2015). Effect of active music therapy and individualized listening to music on dementia: A multicenter randomized controlled trial. *Journal of the American Geriatrics Society, 63*(8), 1534–1539.

*Raglio, A., Bellelli, G., Mazzola, P., Bellandi, D., Giovagnoli, A. R., Farina, E., . . . Trabucchi, M. (2012). Music, music therapy and dementia: A review of literature and the recommendations of the Italian Psychogeriatric Association. *Maturitas, 72*, 305–310.

Raglio, A., Bellelli, G., Traficante, D., Gianotti, M., Ubezio, M C., Gentile, S., . . . Trabucchi, M. (2010). Efficacy of music therapy treatment based on cycles of sessions: A randomised controlled trial. *Aging & Mental Health, 14*(8), 900–904.

Reisberg, B., Monteiro, I., Torossian, C., Auer, S., Shulman, M. B., & Ghimire, S. . . . Xu, J. (2014). The BEHAVE-AD Assessment System: A perspective, a commentary on new findings, and a historical review. *Dementia and Geriatric Cognitive Disorders, 38*(1–2), 89–146. doi:10.1159/000357839.

Ridder, H. M., & Wheeler, B. L. (2015). Music therapy for older adults. In B. L. Wheeler (Ed.), *Music therapy handbook* (pp. 367–378). New York, NY: Guilford Press.

Rolvsjord, R. (2010). *Resource-oriented music therapy in mental health care*. Gilsum, NH: Barcelona.

Seignourel, P. J., Kunik, M. E., Wilson, N., & Stanley, M. (2008). Anxiety in dementia: A critical review. *Clinical Psychology Review, (28)*7, 1071–1082.

Shea, B. J., Grimshaw, J. M., Wells, G. A., Boers, M., Andersson, N., Hamel, C., . . . Bouter, L. M. (2007). Development of AMSTAR: A measurement tool to assess the methodological quality of systematic reviews. *BMC Medical Research Methodology, 7*(10). doi:10.1186/1471-2288-7-10

Shea, B. J., Hamel, C., Wells, G. A., Bouter, L. M., Kristjansson, E., Grimshaw, J., . . . Boers, M. (2009). AMSTAR is a reliable and valid measurement tool to assess the methodological quality of systematic reviews. *Journal of Clinical Epidemiology, 62*, 1013–1020.

Smith, V., Devane, D., Begley, C. M., & Clarke, M. (2011). Methodology in conducting a systematic review of systematic reviews of healthcare interventions. *BMC Medical Research Methodology, 11*(15). doi:10.1186/1471-2288-11-15

Seo, H. J., & Kim, K. U. (2012). Quality assessment of systematic reviews or meta-analyses of nursing interventions conducted by Korean reviewers. *BMC Medical Research Methodology, 12*(129). doi:10.1186/1471-2288-12-129

Stige, B. (2002). *Culture-centered music therapy*. Gilsum, NH: Barcelona.

Tible, O. P., Riese, F., Savaskan, E., & von Gunten, A. (2017). Best practice in the management of behavioral and psychological symptoms of dementia. *Therapeutic Advances in Neurological Disorders, 10*(8), 297–309. doi:10.1177/1756285617712979

*Ueda, T., Suzukamo, Y., Sato, M., & Izumi, S. (2013). Effects of music therapy on behavioral and psychological symptoms of dementia: A systematic review and meta-analysis. *Ageing Research Reviews, 12*, 628–641.

van der Linde, R. M., Stephan, B. C. M., Savva, G. M., Dening, T., & Brayne, C. (2012). Systematic reviews on behavioural and psychological symptoms in the older or demented populations. *Alzheimer's Research & Therapy, 4* (28), 1–23. doi:10.1186/alzrt131

Van der Mussele, S., Le Bastard, N., Saerens, J., Somers, N., Mariën, P., Goeman, J. . . . Engelborghs, S. (2015). Agitation-associated behavioral symptoms in mild cognitive impairment and Alzheimer's dementia. *Aging & Mental Health, 19*(3), 247–257.

*van der Steen, J. T., van Soest-Poortvliet, M. D., van der Wouden, J. C., Bruinsma, M. S., Scholten, R. J. P. M., & Vink, A. C. (2017). Music-based therapeutic interventions for people with dementia [Review]. *Cochrane Database of Systematic Reviews, 5*, Art. No.: CD003477. doi:10.1002/14651858.CD003477.pub3

*Vasionytė, I., & Madison, G. (2013). Musical intervention for patients with dementia: A meta-analysis. *Journal of Clinical Nursing, 22*, 1203–1216, doi:10.1111/jocn12166

Vink, A. C., Zuidersma, M., Jonge, P. D., Zuidema, S. U., & Slaets, J. P. J. (2012). The effect of music therapy compared with general recreational activities in reducing agitation in people with dementia: a randomised controlled trial. *International Journal of Geriatric Psychiatry, 28*, 1031–1038. doi:10.1002/gps.3924

*Wall, M., & Duffy, A. (2010). The effects of music therapy for older people with dementia. *British Journal of Nursing, 19*(2), 108–113.

*Zhang, Y., Cai, J., An, L. Hui, F., Ren, T., Ma, H., & Zhao, Q. (2017). Does music therapy enhance behavioral and cognitive function in elderly dementia patients? A systematic review and meta-analysis. *Ageing Research Reviews, 35*, 1–11.

# 11

# Music Therapy and Skill Sharing to Meet Psychosocial Needs for Persons With Advanced Dementia

*Hanne Mette Ridder and Julie Ørnholt Bøtker*

## Introduction

When music experiences are part of mutual and shared practices in dementia care, several underlying mechanisms and micro processes are at play. The reciprocity that occurs during these musical interactions is essential for meeting the client's psychosocial needs. Our aim with this chapter is to connect the evidence and theories to real life practices and to explore these mechanisms and micro processes. We will thus explain the experiences of a person with dementia, her music therapist and her professional carer at a residential aged care facility. The case narrative features throughout the chapter and is divided into seven parts according to the following themes: (a) listening, (b) singing together, (c) play, (d) carer burden, (e) relationship, (f) safety, and (g) finding the key.

The case is based on a music therapist's (Julie) individual music-therapy sessions with a person with severe dementia (Beth) and Julie's collaboration with a staff member (Lis). Julie took part in a human science research project consisting of a series of three 3-hour workshops for experienced music therapists led by music-therapy researcher Margrete Bach Madsen. The three workshops were audio- and video recorded and transcribed. In this context, Julie told the detailed story about her individual music-therapy sessions with a person, we here name Beth, and about her collaboration with staff. We have shortened, condensed, and anonymized Julie's narrative and added details from her therapy notes. This rich qualitative data was collected as part of the *Person-Attuned Musical Interaction (PAMI) study*, running at Aalborg University from 2015-2019 with funding from the Velux Foundations

(PAMI, 2018; Ridder, Madsen, Anderson-Ingstrup, & McDermott, 2017). Ethics exemption was granted from *Den Videnskabsetiske Komité for Region Nordjylland* and the study was registered at *The Danish Data Protection Agency* through Aalborg University. Throughout the process, we followed and integrated the ethical principles from *The Danish Music Therapist Association* and *The Danish Code of Conduct for Research Integrity*.

The empirical data presented in the case narrative served to provide insight into the complex social phenomenon of person-attuned musical interaction and expand the knowledge of this phenomenon in order to potentially change everyday practices in dementia care. We use the case of Beth in order to explore and explain not only explicit knowledge but also tacit and embedded knowledge from everyday experiences (Nonaka & von Krogh, 2009). Knowledge is constructed to convey *meaning*, and with this said, it is not our intention to create knowledge that brings us *truth*. Rather, we explore how a milieu therapeutic approach to music therapy meets psychosocial needs for persons with advanced dementia and brings knowledge to the interdisciplinary field of care. We explain our theory from an ideographic perspective, and therefore do not claim that our understanding is true in other situations and other cultures of care. However, we may bring this knowledge into a more general understanding of dementia, music therapy, and care. Our aim is to explain a psychosocial and milieu therapeutic approach to music therapy through exploring Beth, Lis, and Julie's interactions.

## Listening

Beth has severe Alzheimer's disease (AD) in addition to chronic obstructive pulmonary disease. She is almost blind (only 3% vision remaining). When Julie meets her, Beth has an MMSE score of 5/30. Beth previously worked as a carer and is not satisfied with the way the carers at her facility do their work, and reacts with sarcasm or anger. Lis has a lot of work to accomplish during her day shift at the residential aged care facility, and it is often disturbing and challenging to get things done while simultaneously dealing with Beth's dementia symptoms.

In this first example (see Box 11.1), Beth's noisy and continuous tapping challenges Lis. Julie, the music therapist, is not able to solve "the problem" but instead sits down and listens to Beth's tapping. Next, Julie joins Beth's

---

**Box 11.1.**

---

*Tack. Tack-tacke. Tack-ke. Tack* . . . Beth's finger nails are tapping rapidly towards the table. Then towards her armrest. Then the table again.

Julie has worked for several years as a music therapist in social psychiatry, but has now started a new position in a residential aged care facility. It is an early afternoon in 2014 and the first time she meets Beth.

Beth is in her late 80s: a small, frail woman with short, fluffy, white hair. Due to failing sight, she is wearing big, dark sunglasses. She sits in the large kitchen/dining room, where Lis, a carer at the unit, is cleaning up. Lis looks at Julie with despair: Beth's tapping and knocking is driving her crazy!

Julie takes a seat next to Beth. She listens. Then joins Beth's tapping. Suddenly she recognizes that the energetic rhythm is similar to one from an old Danish folk tune.

---

tapping, and in this way tries to "be with her" and understand her tapping as an expression of who she is, offering a mode of joint activity.

Music activities and music therapy are considered to be expressions of sound. Sound making is important for engaging in music, but rather than making sound, we want to stress *listening* as a first step. Listening is related to being open and attentive, and to observing, absorbing, learning, and awaiting. Listening implies curiosity and the ability to empathetically take in information and relate to this information. In this sense, listening is where communication starts. Listening sounds easy but can be a rather difficult task (Fromm, 1994). For Lis, listening is challenging as she is busy and under pressure. She does not listen to Beth, but she certainly hears her. The nature of listening is different from the nature of hearing, and even more complex. When the music therapist is taking a "listening" position, she may be hearing or experiencing something very different from what a person under pressure and challenged by tasks is hearing.

By listening, knowledge is assessed and created (Pedersen, 1997). According to Nonaka and von Krogh (2009), knowledge creation is a process where information is assessed and amplified by individuals in an organization (a care facility in our example), and then crystallized, shared, and transferred into the knowledge system in the organization. It is important to explore how knowledge is assessed and amplified between professionals in situations where, for example, one person sees "disturbing" or "annoying" behaviour and the other sees (musical) potential. Contradictory viewpoints of professionals are problematic, unless the workplace culture offers ways of sharing and expanding knowledge. Making use of existing and new tacit and explicit knowledge means enhancing knowledge capacity (Nonaka & von Krogh, 2009). Knowledge may inform ways of being and acting, which is mostly tacit knowledge. Such tacit knowledge further influences how tasks are solved through more explicit social practices.

Beth took part in individual music therapy with Julie for the following year, and Julie was able to share knowledge gained in these sessions with Lis and other carers. For example, Julie could use her knowledge about the concept of agitation to give new insight and knowledge creation, and in this way bridge practices between her and the other professionals. Agitation is common in people with dementia and includes a range of behaviours such as tapping, in the way Beth does, or such as restlessness, pacing, repetitive vocalizations, or aggression. Agitation can cause significant distress, carer burden, and substantial health and social care costs (Livingston et al., 2017). There is thus a need for "new approaches to develop staff skills in understanding and responding to the underlying reasons for individual resident's agitation" (Livingston et al., 2017, p. 171).

## Singing Together

Several studies and meta-reviews have found that music therapy is effective in reducing agitation (see Chapter 12; Abraha et al., 2017; Hsu, Flowerdew, Parker, Fachner, & Odell-Miller, 2015; Pedersen, S., Andersen, Lugo, Andreassen, & Sütterlin, 2017; Ridder, Stige, Quale, & Gold, 2013; Zhang et al., 2017). Beth is almost blind, and therefore easily confused and has difficulty interpreting her environment. Her disturbed sensory perception and cognitive decline make it very challenging for her to participate in social activities, and she is highly dependent on carers. In our attempt to understand Beth's situation, we need to consider her bodily and neurological symptoms, psychological mechanisms, and possibilities for social interaction. British dementia researcher Tom Kitwood was an early advocate for this "biopsychosocial" perspective (Kitwood, 1997). In order to also consider the context and "the relationships between a group of living things and their environment," researchers have suggested the term "ecopsychosocial" (Zeisel, Reisberg, Whitehouse, Woods, & Verheul, 2016), with "eco" representing the Greek word *oikes* for house or household. We use the term *psychosocial* as an integrative term where the psychosocial needs are understood as highly influenced by physical, medical, neurological, and ecological conditions and circumstances. This implies the importance of holistically integrating a broader understanding of dementia symptoms where the expression of behavioural and psychological symptoms could be linked to an expression of "unmet needs" (Kitwood, 1997; Kovach, Noonan, Schlidt, & Wells, 2005).

Kitwood strongly criticized conventional dementia care and opposed treating people with dementia as objects and nonpersons. In his theory of personhood he placed the *person with dementia* at the center, not the disease, and based on observations of practice, he defined a cluster of basic psychological needs in dementia. Due to the cognitive symptoms associated with dementia, "the needs come into the open" (Kitwood, 1997, p. 81). Building on one all-encompassing need for *love*, the main needs are *comfort, attachment, inclusion, occupation,* and *identity* (Kitwood, 1997, p. 81). Kitwood stressed the importance of interpersonal relationships created through positive interactions. With such *positive person work*, it is possible to fulfil the psychological needs and enable the person "to move out of fear, grief and anger, into the domain of positive experience" (Kitwood, 1997, p. 84). Kitwood described "personhood" as the centrality of relationship, the uniqueness of persons, and the fact of our embodiment, and he found parallels to the term personhood in discourses of transcendence, ethics, and social psychology.

Beth understands what Julie communicates to her when she is tapping her rhythm and singing to her (see Box 11.2). It makes sense and gives meaning. At a certain level, Beth is able to relate to the stimuli and engage socially. The song *Rits Rats* is deeply engrained in Beth, and the rhythm is in her body. Even when she is not singing it, the tune is expressed in the way she taps the rhythm. The ability to recognize something when living with visual agnosia and other symptoms of dementia must feel like being thrown a lifebuoy when you are lost at sea. In this way, Beth's ability to recognize the song is "lifesaving." Using music may not work for all, but for many persons with AD, musical memory is well preserved (Jacobsen et al., 2015). For Beth, Julie's rhythmic mirroring and singing is a social interaction that leads to physical and emotional changes.

---

### Box 11.2.

---

*Tack. Tack-tacke. Tack-ke. Tack.* Beth's rhythm perfectly matches the rhythm of the song *Rits Rats filli-Ong gong Gong.* Julie continues tapping with Beth, and then starts singing the lyrics; "*Rits Rats . . . if you want to be my husband, then come, come, come. And the ladies are lifting the skirts . . . Tra-la-la-la.*"

This folk dancing tune is familiar to most Danes, old and young.[a] Immediately Beth joins in singing. She knows all the lyrics and sings the melody exactly in tune. Her rhythmic timing is precise and accurate, and her voice is like a deep sounding whisper. While singing, she rocks vigorously from side to side. She moves her arms and shoulders, and simultaneously makes dancing steps with her feet. A big smile spreads across her face. Julie still clearly remembers this warm meeting.

[a] *Rits Rats*: see the Danish text at: http://www.ugle.dk/rits_rats.html. There are well-known dancing steps to the tune; see: http://roditraditionerne.dk/portfolio/trin-1-rits-rats-med-sideloeb/

---

## Play

For Beth, the song *Rits Rats* becomes the cue or signal to the music-therapy sessions (see Box 11.3). Julie starts with this song each time, and in this way, she gets Beth's attention and focus. Beth is not aware of when their session is scheduled, but in the present moment, Julie's technique is to use *acoustic cuing* by singing this specific song (Ridder, 2011; Ridder & Wheeler, 2015). When Julie has Beth's attention, and her actions are meaningful to Beth, they are able to attune to each other and to play. Julie slightly changes the musical elements (tempo, timbre, and pitch) in the song. A tension is built up, but then she comes back and keeps to the structure of the song. This is the same behaviour we see in play. In play, we modify the well-known but maintain the recognizability. In mutual play, we share experiences about ourselves and creatively express the spontaneous activity of our personality (see Trevarthen, 2016). Play may happen with bodily and vocal expressions and actions and "is a means of mastering the gap between the inner and outer world" (Hart, 2011, p. 95). Through play, children learn to modulate and control emotion and to develop the capacity for symbolic action (Hart, 2011).

---

**Box 11.3.**

---

After repeating the first song (*Rits Rats*) a couple of times, Julie continues with another song. Beth joins her. Julie is impressed by how many of these old Danish tunes Beth knows. She feels how she and Beth connect in a positive and warm way when singing together.

Over the next year, Julie continues with regular individual music-therapy sessions with Beth; in the kitchen, in the lounge, or in Beth's living room. They start every time with *Rits Rats*. Beth laughs loudly when Julie increases the tempo of the song, slows down, or suddenly pauses while anticipating her response. The song is a stable and well-known starting point when they meet, but can then be changed in various playful ways without losing its recognizable form. After this, other songs follow. Beth smiles, laughs, sings, moves, dances, and shows a lot of humor.

---

In his research, psychologist Daniel Stern carried out a large number of micro-analyses of the duration, sequential patterning, rate and rhythmicity in early mother-infant play (Stern, 1971). From this early dyadic playing, Stern and colleagues reasoned that infants must have specific timing mechanisms for allowing them to keep track of the mother's rhythmic social behaviours (Beebe, 2017). With such timing, both mother and infant are able to attune to each other. The infant keeps the mother attuned by responding with vocalizations and smiles. Smiling, laughing, and humor are responses to playful actions. Such responses express the joy of playing, but also keeps playmates engaged in joint action.

Play is a primary element for building culture (Huizinga, 1955) and may represent two different forms; "ludus," where play is role-structured activity often including competitive behaviour, or "paidia" which is free improvisation and the channeling of free energy (Caillois, 1958). In a review of the literature, researchers from the Netherlands found that play experiences that involved *sensation, relaxation*, and *reminiscence* may be relevant and meaningful for persons in advanced stages of AD (Anderiesen, Scherder, Goossens, Visch, & Eggermont, 2015).

Beth and Julie play in the form of "paidia" which has spontaneous and improvisational characteristics. This leads us back to Daniel Stern, who emphasized the role of *improvisation* in music therapy in his explanation of vitality forms and intersubjectivity (Stern, 2010). Vitality forms can be explained as the dynamics of movements and how they are performed. According to Stern, "the basic methods in improvisation music therapy all require the use of vitality forms to share or interchange experience" (Stern, 2010, p. 140). In order to outline the details of improvisation, Stern referred to the work on improvisation in music therapy by Tony Wigram (2004). Wigram described the most commonly used music-therapy improvisation methods, which in summary are: (a) mirroring, (b) matching, (c) empathic improvisation, (d) grounding, (e) dialoguing, and (f) accompanying (Stern, 2010, p. 139; Wigram, 2004), and which Julie uses with Beth. In our example, Julie *mirrors* Beth's tapping rhythmically, but also *matches* the dynamics in Beth's movements or in her vitality forms. She improvises a response to Beth that *empathetically* reflects Beth's emotional state and her "way of being" at that moment. She creates a stable musical "anchor" with the well-known songs and her accompaniment on the guitar, and in this way provides a dynamic framework to *ground* their interaction. She is aware of the smooth turn taking (*dialoguing*) between her and Beth that allows them to be mutually adjusted to the timing of each other. Finally, Julie *accompanies* Beth's musical expressions, be it tapping, moving, humming, or singing, in a way where Beth is the soloist. Julie's accompaniment is simple, repetitious, and sensitive to the pauses and developments in Beth's expression (Stern, 2010, p. 140; Wigram, 2004). Even if Beth's musical inputs are short and contain only a few musical elements, they are enough for Julie to attune to her. The difference between imitation and attunement is that imitation relates to external behaviour, whereas "affective attunement mediates internal states" (Hart, 2011, p. 24). This demands a competence from the music therapist to be attentive to the inner states of the person with dementia.

## Carer Burden

Being able to engage in attuned interaction demands from the carer or the therapist that they interact with reciprocity and mutuality. They have to use their own bodily and musical expressions to create such intersubjective interaction. Therefore, the person with dementia has to rely on the professional's ability to listen, engage, play, and improvise if attunement is to occur. From play therapy we know that play gives us a safe distance to threatening issues, but at the same time, we also need to feel safe to play (Kestly, 2014). If the

---

**Box 11.4.**

---

Julie informs staff about which songs she sings with Beth and how Beth reacts with vitality, playfulness, and engagement. The carers tell another story about Beth. Her endless tapping is disturbing for them, and helping her with personal care is challenging, as she is noncompliant.

---

professional is to build up a safe setting, he/she must not only be able to listen, but also to attune with the other. Attunement will only happen if the professional him/herself feels safe and is able to mediate safety (Ridder, 2017).

Julie is able to attune to Beth in individual music therapy, but Lis only sees Beth when she is engaged in her role as her carer, and in situations where Beth feels unsafe and not able to interact. Beth's actions seem incomprehensible to Lis (see Box 11.4). Even if Julie tells Lis about Beth's large repertoire of songs and her ability to engage and interact in various ways, Lis is not able to use this information. The knowledge Julie has about Beth seems irrelevant to Lis's work. She is thrilled and enthusiastic about the way Julie engages Beth and other residents musically, but she is not aware of how music therapy may play a role for attuned interactions in a broader sense. For Julie, who has previously worked closely together with professionals in another context, this communication difficulty is frustrating as her attempts to share knowledge with Lis seem to not yield results.

## Relationship

In her music-therapy work at the care facility, Julie conducts various therapeutic music interventions, such as group singing or dance/movement. However, it is clear that Beth benefits most from the short individual sessions where Julie attunes to Beth's needs, and a relationship builds up over time. Julie is informed by Beth's life history and with time becomes aware of her typical ways of interacting within the music. The therapeutic relationship between Beth and Julie allows for play and intersubjective meetings where Beth's psychosocial needs are met in the present moment.

In the sessions with Julie, Beth is smiling and is socially engaged (see Box 11.5), but Lis still has severe problems in managing Beth's agitated

---

**Box 11.5.**

---

Beth loses weight rapidly as symptoms of dementia become more severe (MMSE = 0). She does not want to get out of bed. She resists wearing her dentures, which makes her face shrink in even more. For their music-therapy sessions, Julie sits at Beth's bedside. When she sings *Rits Rats*, Beth bobs her feet up and down, conducts the music with her arms in the air and sometimes sings a little. She often smiles and laughs. Julie feels there is a connection between them and that she is getting to know Beth.

---

behaviours, such as making noise and resisting daily care (washing and dressing). The music therapy seems to work only in the moment and not to have any carryover effects. For Julie it is challenging to transfer the positive effects from the individual music therapy to daily care, but finally the breakthrough came.

## Safety

What is magical about the example in Box 11.6 is not the music itself, but that the combination of music used in a planned way with the actions of the carers creates safety. With help of the music, Beth is able to move out of fear and anger, and to move into the domain of positive experience. A situation that initially makes Beth feel highly confused and uncomfortable, is transformed into a safe situation, which even makes it possible for the carers to wash her. A difficult situation is changed to a situation where basic psychosocial needs are met. The music therapist is aware of how to make Beth feel safe, and she and the carers attune their interactions to Beth's psychosocial needs. In this way actions related to a task (and a very challenging one) are transformed to meaningful attuned interactions. Even if Julie uses the same song over and over, she makes use of dyadic playing, empathic improvisation, and affective attunement. She does not mirror or match Beth's fear and anger, but attunes to Beth's need for safety and comfort.

## Box 11.6.

Lis sits on a chair in the corridor outside Beth's living room. She looks exhausted. Julie, who is passing by, hears Beth screaming. Lis tells how Beth is spitting and scratching. Beth and her bed are smeared with feces, but she does not want anybody to wash her. She is agitated and aggressive. They need three carers to help Beth and now it is Lis's turn to take a break. When Julie suggests that she sings *Rits Rats*, Lis looks resignedly at her. Julie understands that her idea sounds stupid and worthless to Lis. When Julie then suggests that she will join them and sing, Lis is reluctant and explains that it is really quite disgusting in there. "Well, let's give it a try," Julie suggests.

While the second carer is taking a break, Lis and the third carer continue to clean up Beth who still lies in her bed. Julie has the guitar with her and starts singing *Rits Rats* at the foot of the bed. She sings in exactly the same way as she sings with Beth in music-therapy sessions. Immediately she sees Beth's feet moving to the beat, then her arms conducting the music. Then Beth joins in humming and the screaming stops. The two carers gently remove Beth's clothes and use washcloths to clean her. They look stunned at this sudden change in Beth's behaviour and manage to help Beth to sit up, to put her legs to the floor, and then help her stand up. They see the opportunity of actually getting Beth to her bathroom for a shower, and one of the carers tell her in a soft voice, "Come Beth, let's go to the bathroom." Beth starts walking and the two carers support her unsteady walking by holding one arm each. Julie follows close by, never stopping the music in order to keep Beth feeling safe.

## Finding the Key

Julie conducted regular music-therapy sessions with Beth for 1 year, and in this period, Beth's deterioration was evident. She was at the end of a long life and dealing with severe dementia and loss of sight. For Lis, the carer, it was very challenging. She did not find the key to understanding Beth, not until Julie had a chance to share her way of engaging with Beth directly in a critical moment. Until the end, Beth enjoyed music and clearly responded to it (see Box 11.7). The music made it possible for her to interact with Julie.

## Box 11.7.

Julie carefully matches Beth's walking pace, and is aware of her tempo and energy. Beth stands up and taps the beat at the sink in her *Rits Rats* rhythm while the carers give her a shower. After more than 20 times with *Rits Rats*, Julie is more than ready to try another song. She understands how staff must have the same reaction to the endless tapping. She has a need to hear something else; to change the soundscape. Julie shifts to a song in the same genre and immediately sees a change. Beth looks baffled at first, then she becomes angry. Julie gently resumes with *Rits Rats*.

Soon after, Beth sits quietly in the dining room, clean, dressed and with her breakfast. Julie now fades out her singing and guitar playing so she can leave Beth. She makes a clear ending by telling her; "Well, Beth, have a nice day." Promptly Beth answers, "Yes, let's hope so." Julie and staff look at each other, all surprised at Beth's clear and appropriate response.

This moment is a turning point for Julie's collaboration with Lis. Finally, she feels, Lis understands *how* she can make use of music and appreciates Julie's music-therapy competencies, not only for the benefit of the residents but also for her own benefits in improving her role as a carer. Lis tells Julie that she has started to sing with Beth, and sometimes also with other residents while doing her caring duties, and that it often helps with their compliance.

*Tack. Tack.* Beth's tapping stopped 12 days after. When her coffin was carried down the corridor and out of the care facility, Julie was in the dining room with a group of residents. Their goodbye was singing Beth's favorite lullaby.

Beth, however, was selective about which music she wanted, and Julie was aware of how to meet her and to build up a session, create safety, and facilitate moments of play and attunement.

Carers have a very difficult task when assisting persons with advanced dementia with actions that are sometimes incomprehensible to them. When Beth was not able to understand the intentions of the carers, she became frightened and angry. From her perspective, they seemed to do things to her that she found unpleasant and threatening. Julie learned that all caring acts had to be carried out in a communicative way. She later explained in the workshop: "it became really clear to me (that) music . . . suddenly can open doors." For Lis, the key to understanding Beth came at a critical moment with the assistance of Julie through music therapy (see Box 11.7). The music made it possible for Beth to interact with Julie and Lis. Julie was aware of how to meet Beth and to create safety and facilitate moments of play and attunement.

## Discussion

We have presented a case example where the music therapist (Julie) learned which person-attuned musical interactions were important for building up a therapeutic relationship with an aged care facility resident with advanced AD (Beth). The music therapist mediated safety and attuned to the resident by engaging in clear rhythmic behaviours and then transferred these behaviours to social behaviours. Rhythmic timing is intrinsic in human communication and may expand to spontaneous and playful ways of improvising. This is such an integrated part of human communication that we are not aware of it. Music therapists are trained to work with and to explicate these tacit micro processes, but for carers, such knowledge may seem irrelevant until they experience *how* it works—not only in scheduled therapy sessions, but in daily care. By working together, as in the example of Julie and Lis (the carer) when bathing Beth, knowledge is implicitly shared. In the case example, the carer was part of the process when the music therapist used music to interact in a way that regulated the dynamic movements of the person with dementia and led to social engagement. The care situation can be transformed through music engagement from being an unpleasant and anxiety provoking experience, to being interactive and pleasant. For people with severe cognitive impairment associated with advanced dementia, it is crucial to "speak to the senses" and be aware of sensory integration and nonverbal language. Music, touch, timing and tone of voice are important ways of communicating in the context of an absence of verbal communication.

Individual music-therapy sessions are important for creating moments where it is possible to attune to the person. This does not necessarily happen behind closed doors. Sometimes it is essential for building up trust that the sessions take place in the same room every time, with a clear structure and with closed doors to hinder disturbances. Sometimes it is safe enough to have this interaction in an open setting and with the presence of other people. An individual session may even be planned as part of a group session, if this is the best way to meet psychosocial needs for the person. When we suppose that knowledge creation and knowledge sharing happens in the interactive field, it is important to consider when individual sessions benefit from including other persons, such as carers, relatives or volunteers. Sharing knowledge with other persons may lead to new insights and new ways of acting and interacting and in this way changing the culture of care and the reflections about daily social practices.

**Carers' use of person-attuned musical interactions.** Singing a well-known song is something most people are able to do, but for various reasons many people are not comfortable when singing to others. Our voice is closely connected to our identity, and to expose it by singing may lead to insecurity. Singing in a spontaneous and improvisational way may be even more difficult, although most preschool children do it with joy and in a way that is perfectly free from the social constraints that later seems to dampen such behaviour. With this in mind, it is too demanding to expect from carers that they spontaneously sing and use music in their caregiving roles. They must have a chance to build such competencies for use of music in caregiving through guidance and learning. The music therapist may function as a role model and teacher, and in this way knowledge is shared through situated (Lave & Wenger, 1991) and experimental learning (Dewey, 1934). The next step is that knowledge is transferred to the entire organization (residential aged care facility) by explicit and reflective processes. The skills required to listen, regulate, play, and attune through musical interaction are learned by experiencing how it works and by reflecting and experimenting.

The collaboration between music therapist and carers does not only derive from scheduled work. It must be possible also to draw on the competencies of the music therapist in acute situations, learning by doing, and afterwards reflecting and evaluating, for new knowledge to be integrated in future interactions and interventions.

**From knowledge creation to skill sharing.** Transforming tacit knowledge to explicit knowing is difficult, as is skill sharing between different professions (McDermott, Ridder, Baker, Wosch, Ray, and Stige, 2018). We want to stress that it is not enough to simply tell a carer to use music and to sing. In our example, Lis (the carer) could see and observe that Beth (the person with dementia) enjoyed music, but Lis could not integrate it as a tool in her own work as a carer. There is a need for knowledge transfer, and although many carers intuitively use musical and attuned interactions, consistent use of attuned music interaction in caregiving will only happen when leaders and politicians prioritize knowledge creation as part of practice learning. The proficiency of dementia care is influenced by the way professionals collaborate and learn from each other and have a chance to share knowledge, reflections, and skills. The person with dementia relies on the professional's competencies to listen, engage, play, and improvise if attunement is to occur. As it is now, such competencies can be found in individual carers, but are not competencies that are generally nurtured or demanded.

## Conclusion

Currently, only a few people with dementia are lucky enough to live in a care home where the professionals have the required competencies to use music in an attuned way. All people with dementia should be able to expect a high level of interpersonal competencies from their professional carers. There is a need to gain knowledge about *how* psychosocial needs are met in persons with dementia. A first step is to increase knowledge about how to use person-attuned musical interactions in the interdisciplinary field. For persons with advanced dementia and no verbal language, affective attunement is a way to mediate internal states. Being able to engage in attuned interaction demands from the carer or the therapist that they interact with reciprocity and mutuality, using their own bodily and musical expressions. Further, this demands an organization and leaders who understand and foster knowledge sharing.

## References

Abraha, I., Rimland, J. M., Trotta, F. M., Dell'Aquila, G., Cruz-Jentoft, A., Petrovic, M., . . . Cherubini, A. (2017). Systematic review of systematic reviews of non-pharmacological interventions to treat behavioural disturbances in older patients with dementia. The SENATOR-OnTop series. *BMJ Open, 7*(3), 1–28.

Anderiesen, H., Scherder, E. J. A., Goossens, R., Visch, V., & Eggermont, L. (2015). Play experiences for people with Alzheimer 's disease. *International Journal of Design, 9*(2), 155–165.

Beebe, B. (2017). Daniel Stern: Microanalysis and the empirical infant research foundations. *Psychoanalytic Inquiry, 37*(4), 228–241.

Caillois, R. (1958). *Man, play, and games.* Champaign, IL: University of Illinois Press.

Dewey, J. (1934). *Art as experience.* New York: Berkley Publishing Group.

Fromm, E. (1994). *The art of listening.* New York: Continuum International Publishing Group.

Hart, S. (2011). *The impact of attachment. Developmental neuroaffective psychology.* London: Norton.

Hsu, M. H., Flowerdew, R., Parker, M., Fachner, J., & Odell-Miller, H. (2015). Individual music therapy for managing neuropsychiatric symptoms for people with dementia and their carers: A cluster randomised controlled feasibility study. *BMC Geriatrics, 15*(1), 1–20.

Huizinga, J. (1955). *Homo ludens. A study of the play element in culture.* Boston, MA: Beacon Press.

Jacobsen, J. H., Stelzer, J., Fritz, T. H., Chételat, G., La Joie, R., & Turner, R. (2015). Why musical memory can be preserved in advanced Alzheimer's disease. *Brain, 138*(8), 2438–2450.

Kestly, T. A. (2014). *The interpersonal neurobiology of play: Brain-building interventions for emotional well-being.* London: Norton.

Kitwood, T. (1997). *Dementia reconsidered. The person comes first.* Buckingham: Open University Press.

Kovach, C. R., Noonan, P. E., Schlidt, A. M., & Wells, T. (2005). A model of consequences of need-driven, dementia-compromised behavior. *Journal of Nursing Scholarship, 37*(2), 134–140.

Lave, J., & Wenger, E. (1991). *Situated learning: Legitimate peripheral participation.* Cambridge: Cambridge University Press.

Livingston, G., Barber, J., Marston, L., Rapaport, P., Livingston, D., Cousins, S., . . . Cooper, C. (2017). Prevalence of and associations with agitation in residents with dementia living in care homes: MARQUE cross-sectional study. *British Journal of Psychiatry Open, 3*(4), 171–178.

McDermott, O., Ridder, H. M., Baker, F. A., Wosch, T., Ray, K., & Stige, B. (2018). Indirect music therapy practice and skill-sharing in dementia care. *Journal of Music Therapy, 55*(3), 255–279. doi:https://doi.org/10.1093/jmt/thy012

Nonaka, I., & von Krogh, G. (2009). Tacit knowledge and knowledge conversion: Controversy and advancement in organizational knowledge creation theory. *Organization Science, 20*(3), 635–652.

Pedersen, I. N. (1997). The music therapist's listening perspectives as source of information in improvised musical duets with grown-up, psychiatric patients, suffering from schizophrenia. *Nordic Journal of Music Therapy, 6*(2), 98–111.

Pedersen, S. K. A., Andersen, P. N., Lugo, R. G., Andreassen, M., & Sütterlin, S. (2017). Effects of music on agitation in dementia: A meta-analysis. *Frontiers in Psychology, 8,* (May, Article 742), 1–10.

Person-Attuned Musical Interaction. (2018). *Person-attuned musical interaction in dementia care.* Retrieved from https://www.musictherapy.aau.dk/pami/

Ridder, H. M. (2011). How can singing in music therapy influence social engagement for people with dementia. Insights from the polyvagal theory. In F. Baker & S. Uhlig (Eds.). *Voicework in music therapy. Research and practice* (pp. 130–146). London: Jessica Kingsley.

Ridder, H. M. (2017). Partners in care: A psychosocial approach to music therapy and dementia. In S. L. Jacobsen & G. Thompson (Eds.), *Music therapy with families: Therapeutic approaches and theoretical perspectives* (pp. 269–289). London: Jessica Kingsley.

Ridder, H. M., Madsen, M. B., Anderson-Ingstrup, J., & McDermott, O. (2017). The development of person attuned musical interaction (PAMI) for people with dementia. *Music Therapy Today: WFMT Online Journal, 13*(Special Issue), 136–137.

Ridder, H. M., Stige, B., Quale, L. G., & Gold, C. (2013). Individual music therapy for agitation in dementia : An exploratory randomized controlled trial. *Aging & Mental Health, 17*(6), 667–678.

Ridder, H. M., & Wheeler, B. L. (2015). Music therapy for older adults. In B. L. Wheeler (Ed.), *Music therapy handbook* (pp. 367–378). New York: The Guildford Press.

Stern, D. N. (1971). A micro-analysis of mother-infant interaction: Behavior regulating social contact between a mother and her 3 1/2-month-old twins. *Journal of the American Academy of Child Psychiatry, 10*(3), 501–517.

Stern, D. N. (2010). *Forms of vitality. Exploring dynamic experience in psychology, the arts, psychotherapy, and development.* Oxford: Oxford University Press.

Trevarthen, C. (2016). From the intrinsic motive pulse of infant actions to the life time of cultural meanings. In B. Mölder, V. Arstila, & P. Øhrstrøm (Eds.), *Philosophy and psychology of time* (pp. 225–265). New York: Springer.

Wigram, T. (2004). *Improvisation: Methods and techniques for music therapy clinicians, educators, and students.* London: Jessica Kingsley.

Zeisel, J., Reisberg, B., Whitehouse, P., Woods, R., & Verheul. A. (2016). Ecopsychosocial Interventions in cognitive decline and dementia: A new terminology and a new paradigm. *American Journal of Alzheimer's Disease and Other Dementias, 31*(6), 502–507.

Zhang, Y., Cai, J., An, L., Hui, F., Ren, T., Ma, H., & Zhao, Q. (2017). Does music therapy enhance behavioral and cognitive function in elderly dementia patients? A systematic review and meta-analysis. *Ageing Research Reviews, 35*, 1–11.

# 12

# Music Interventions
# for Advanced Dementia

## Needs and Clinical Interventions Identified From a Narrative Synthesis Systematic Review

*Melissa Mercadal-Brotons*

The number of people living with dementia worldwide is currently estimated at 35.6 million (Alzheimer's Disease International & World Health Organization, 2017). This number will double by 2030 and more than triple by 2050 (Prince et al., 2016). The total number of new cases of dementia each year worldwide is nearly 7.7 million, implying one new case every 4 seconds (Alzheimer's Disease International & World Health Organization, 2017). Dementia is an overwhelming diagnosis not only for the people with the condition, but also for their caregivers and families, as it is one of the major causes of disability and dependency among older people. The high global prevalence, and the economic impact of dementia on families, caregivers, and communities, present significant public health challenges (Prince et al., 2015).

Dementia is a syndrome due to chronic or progressive brain disease in which there is disturbance of multiple higher cortical functions, including memory, thinking, orientation, comprehension, calculation, learning capacity, language, and judgement (American Psychiatric Association, 2013). Alzheimer's disease (AD) is the most prevalent type of dementia contributing to 60–70% of the cases (Alzheimer's Disease International & World Health Organization, 2017), but there are other types. The boundaries between subtypes of dementia are indistinct, and mixed forms often coexist (Zheng, Vinters, Mack, Weiner, & Chui, 2016).

Due to the degenerative nature of dementia, impairments in memory, thinking, and behavior continue to advance as the disease progresses. Although dementia affects each person in a different way, there are three main stages: (a) early or mild stage, (b) middle or moderate stage, and (c) late or advanced stage (Fisher Center for Alzheimer's Research Foundation, 2017), and each of these stages is characterized by a series of symptoms. This review focuses on music intervention for people in the late or advanced stage of dementia who present with nearly total dependence and inactivity. Cognitive deficits are serious and the physical side of the disease becomes more noticeable (Alzheimer's Society, 2018). Specifically, symptoms in the later stages of dementia include (a) severe memory loss, (b) communication problems, (c) inability to recognize relatives, friends, and familiar objects, (d) increasing need for assisted self-care, (e) incontinence, (f) frailty, (g) difficulties with mobility, (h) changes in behavior such as aggression and agitation, and (i) disorientation to time and place.

No treatments are currently available to cure the progressive course of dementia, although numerous new therapies are being investigated in various stages of clinical trials (Cooper, 2014). A number of studies have examined pharmacological treatments in dementia, and future research will continue to explore this avenue (Dyer, Harrison, Laver, Whitehead, & Crotty, 2018). However, the adverse and limited effects of drugs have boosted the exploration of nonpharmacological therapies and their efficacy in helping people with dementia and their caregivers to manage the symptoms and in potentially slowing the progression of the disease (Ballard, Khan, Clack, & Corbett, 2011). Nonpharmacological therapies, including music-based interventions, have become increasingly popular and have proven to be effective in managing a variety of symptoms (including agitation) that are common in the middle or later stages of dementia (Abraha et al., 2017). These therapies don't have unpleasant side effects and thus appear to be a beneficial alternative to pharmacological treatment.

It is widely recognized that musical abilities may outlast other faculties in advanced dementia and may become almost the sole means of communication between patient and carer (Jacobsen et al., 2015). Musical memory is considered to be partly independent from other memory systems such as episodic, semantic, and working memory (Boso, Politi, Barale, & Enzo, 2006). In AD and other types of dementia such as frontotemporal dementia, musical memory is surprisingly robust since the musical memory region features among the lowest grey matter atrophy and hypometabolism values of the entire brain (Lin et al., 2016). Further, the ventral presupplementary motor area and caudal anterior cingulate gyrus are among the last regions to degenerate during AD (Janata, 2009). Music therapy (MT) appears to be an effective approach to treat a variety of symptoms in late-stage dementia, providing a low-cost alternative to medication to reduce behavioral disturbances with fewer negative side effects (Patel, Perera, Pendleton, Richman, & Majumdar, 2014; Raglio, Filippi, Bellandi, & Stramba-Badiale, 2014; Sjogren, Lindkvist, Sandman, Zingmark, & Edvardsson, 2013).

In addition, activity theory (Engeström, Mettinen, & Punamaki, 1999) proposes that individuals who are engaged in the world around them experience increased levels of psychological well-being and improved quality of life (QoL). In recent years, this view has become very influential in the field of dementia care. Therapeutic activities provide physical, mental, and emotional stimulation to engender a sense of meaning, accomplishment, and belonging (Phineey, Chaudhury, & O'Connor, 2007). Currently, several categories of activity-based therapeutic interventions are available, including various art activities and creative art therapies (Gross, Danilova, Vandehey, & Diekhoff, 2015). Caregivers can benefit indirectly from art therapies as well, as seen by improvements in their attitudes and morale (Karli, Young, & Dash, 2017).

Systematically evaluating the effects of music interventions in dementia care is important for assessing the efficacy of various interventional approaches. Up to the beginning of the 21st century, the majority of MT research was conducted with individuals in early and middle stages of dementia, particularly AD. In the late 1990s, Clair (1996) and Clair and Ebberts (1997) carried out studies with people in late-stage dementia, and showed that even in the late stages of dementia people can continue participating in meaningful music-based activities using rhythm and dancing. Since then, progressively more studies have been conducted with individuals who are in the late stages of dementia (Olley & Morales, 2017).

In this chapter the scientific literature, published from 2000–2017 in English and Spanish, concerning the use of music and MT with people in the late stages of dementia of all types will be reviewed with the following objectives:

1. To identify the symptoms of people in the late stages of dementia.
2. To identify the domains and therapeutic objectives addressed in music and MT interventions for people in late stages of dementia.
3. To highlight specific methodological aspects of music and MT applications with people in late stages of dementia: specific music

techniques, dosage of sessions (number, frequency, and duration of sessions), therapeutic setting (group vs. individual sessions).

4. To provide evidence-based recommendations to guide music therapists in the selection of appropriate MT interventions and assist caregivers to make informed choices about appropriate music to use with their care recipients in late stages of dementia.

Exploring these aspects is clinically relevant and has important implications for music therapists as well as for health service planning.

## Method

Computerized healthcare electronic databases were searched, including Medline, PsycINFO, CINAHL, Scopus, Academic Search Complete, and Science Direct, using the following combined keywords: music therapy, music, advanced dementia, severe dementia and late stage dementia.

## Inclusion Criteria

Only original articles that evaluated music-based and MT interventions were included. These were primary research studies which either used quantitative or qualitative research methodologies. The quality of research methodologies was not evaluated in this review since it aimed to include as many sources as possible. Systematic reviews were not included in the analysis, only primary studies. Specific criteria for inclusion were the following:

- Studies needed to include people in late stages of dementia and this had to be specified in the article.
- Music was one of the intervention stimuli regardless of how it was implemented (e.g., singing, playing instruments, listening).
- Study results reported on the impact of music or MT interventions on a functional area: physical, cognitive, social, emotional, and the spiritual domain.
- Scientific articles were published in English or Spanish.
- Articles were published 2000–2017.

## Data Extraction and Analysis

Studies which included specific music or MT interventions are synthesized in Table 12.1 according to the following characteristics: author, year, country, sample characteristics (sample size and stage of the disease), area(s) addressed in the intervention, type of music intervention, individual/group intervention, dosage (number, frequency, and length of sessions), and interventionist. Findings were then divided according to the domains addressed: physiological; cognitive; social/emotional; behavioral-psychological symptoms of dementia with a special focus on agitation, apathy, social withdrawal; and quality of life. A further section on the training of caregivers to deliver music-based interventions was also included. A narrative synthesis approach to analysis was taken, where primarily words and text are used to examine, summarize, and explain the findings. Please note that many studies address several areas of intervention.

## Results

Twenty-five articles were included in this review, with a total of 795 participants in various stages of dementia and 39 caregivers. Most studies were characterized by small sample sizes and high drop-out rates. Fifteen studies defined and identified the late stages of dementia through scores in standardized tests. Ten of these used the Mini-Mental State Examination (MMSE), and the others used scales such as Reisberg's Global Deterioration Scale (GDS), Reisberg's Functional Screening Test (FAST), and the Revised Hasegawa Dementia Scales (HDS-R). Regarding intervention dosage, 22 of the studies reviewed were specific about dosage of the music intervention and the range was quite diverse. The total number of sessions ranged from 1–30, and the frequency varied from daily intervention to one session per week for several weeks. The duration of the interventions also varied between studies, ranging from 6 to 90 minutes, although a 30-minute intervention length was most common.

It is important to differentiate two main types of studies: (a) studies that evaluated a specific music intervention, implemented by a professional music therapist, addressing specific domains as therapeutic objectives, and (b) studies that compared different types of stimuli, among them music, usually led by professionals or researchers other than music therapists.

Table 12.1. Summary of Studies

| Author | Year | Country | N/Stage Dementia | Areas Addressed | Music Intervention | Individual/ Group | # Sessions | Frequency of Sessions | Length of Sessions | Professional |
|---|---|---|---|---|---|---|---|---|---|---|
| | | | | | PHYSIOLOGICAL | | | | | |
| Takahashi et al. | 2006 | Japan | 43/ Moderate-severe | Physiological measures: cortisol level in saliva and blood pressure | Active reminiscence MT group: exercise to music, singing, playing in a concert | Group | NS | 2 years of weekly MT | 60 min. | MTx |
| | | | | | COGNITIVE | | | | | |
| Van de Winckel et al. | 2004 | Belgium | 25/. Moderate-severe | -Cognition -Behavior | -Music-based dance therapy -Recorded music | Group | NS | Daily for 3 months | 30 min. | PT |
| Belgrave | 2009 | USA | 9/ Late stage | -Alert behavior | -Expressive touch + singing -Instrumental touch + singing | Individual | 9 | 2/week | 30 min. | MTx |
| Clements-Cortés et al. | 2016 | Canada | 18/ Moderate-advanced | -Alertness -Cognition -Short-term memory | RSS at 40Hz | Individual | 13 | 2/week (8) | 30 min. | MTX |
| Dassa | 2014 | Israel | 6/Middle-late stage | -Communication -Conversation | Singing familiar songs | Group | 8 | 2/week (4) | 45 min. | MTx |

(continued)

Table 12.1. Continued

| Author | Year | Country | N/Stage Dementia | Areas Addressed | Music Intervention | Individual/ Group | # Sessions | Frequency of Sessions | Length of Sessions | Professional |
|---|---|---|---|---|---|---|---|---|---|---|
| | | | | | SOCIAL/EMOTIONAL | | | | | |
| Ridder & Gummersen | 2015 | Denmark | 1/ Moderate-severe | -Communication and dialogue | -Singing -Listening -Playing instruments -Improvising | Individual | 14 | 1/week | NS | MTx |
| Schall et al. | 2015 | Germany | 9/Advanced | -Communication -Expression of emotions, -Situational well-being | -Active MT -Receptive MT | Individual | 20–27 | Weekly over 6 months | 23–39 min. | MTx |
| | | | | | BEHAVIORAL & PSYCHOLOGICAL SYMPTOMS OF DEMENTIA (BPSD) | | | | | |
| Gerdner | 2005 | USA | 8/Severe-very severe | -Agitation | -Listening to individualized preferred music | Individual | NS | Daily (8 weeks) | 30 min. | CNAs and family |
| Holmes et al. | 2006 | UK | 32/ Moderate-severe | -Apathy | -Live interactive music: musicians playing -Passive prerecorded music or silence | NS | 1 | — | 1.5 h. | Musicians |
| Tuet et al. | 2006 | China | 16/ Moderate-severe | -Agitation | -Active music: singing, playing instruments, exercise with music | Group | 9 | 3/week | 45 min. | OT and 2 assistants |

| Author | Year | Country | N/Severity | Outcomes | Intervention | Setting | N | Frequency | Duration | Provider |
|---|---|---|---|---|---|---|---|---|---|---|
| Svansdottir et al. | 2007 | Iceland | 38/ Moderate-severe | -Activity disturbances -Aggression -Anxiety | -Active MT: singing, instrument playing, improvisation; | Group | 18 | 3/week (6) | 30 min. | MTx |
| Raglio et al | 2008 | Italy | 59/ Moderate-severe | -BPSD: delusions, agitation, apathy, anxiety | -MT vs. entertainment. -Active MT: playing musical instruments | NS | 30 | 16 weeks | 30 min. | MTx |
| Raglio et al. | 2010 | Italy | 60/Severe | -Behavioral disturbances | -Active MT sessions: improvisation using musical instruments. | Group | 36 | 3/week | 30 min. | MTx |
| Sakamoto et al. | 2013 | Japan | 39/Severe | -Emotional response -Stress level -BPSD | -Passive/interactive music interventions | Individual | 10 | 1/week (10) | 30 min. | 2 MTx 4 OT, 6 RN |
| Ridder et al. | 2013 | Denmark | 42/ Moderate severe | -Agitation -QoL | -Individualized MT: singing, listening and dancing | Individual | 12 | Biweekly (6) | 30 min. approx. | MTx |
| Gold | 2014 | UK | 9/Severe | -Mood -Behavior | -Active MT: Singing, playing instruments, improvising | Group | 16? | 1/week (4 months) | 45-60 min | MTx |
| Raglio et al. | 2015 | Italy | 120/ Moderate-severe stages | -Depression -Anxiety -Agitation -Apathy -QoL | -Music listening -Active MT: singing, playing instruments, improvisation | Individual | 20 | 2/week | 30-min | MTx |

(continued)

Table 12.1. Continued

| Author | Year | Country | N/Stage Dementia | Areas Addressed | Music Intervention | Individual/ Group | # Sessions | Frequency of Sessions | Length of Sessions | Professional |
|---|---|---|---|---|---|---|---|---|---|---|
| Ray et al. | 2015 | USA | 132/ Moderate-severe | -Depression -Agitation -Wandering | -Singing -Instrument playing -Music and movement | Groups | 6 | 3/week (2) | 15-60-min. | MTx |
| **QUALITY OF LIFE** | | | | | | | | | | |
| Van der Vleuten | 2012 | The Netherlands | 45/ Moderate-severe | -QoL: Participation mental well-being -Emotions -Communication | -Intimate live music performances | Group | 1 | NS | 45 min | Professional singers |
| Solé et al.[1] | 2014 | SPAIN | 16 /Mild-moderate-severe | -QoL -Affect -Participation | -Active MT: singing, playing musical instruments, moving to music | Group | 12 | 1/weekly (12) | 45–60 min. | MTx |
| Rubbi et al | 2016 | Italy | 32/ Moderate-severe | -QoL | -Video MT -Video projections of: folk traditions music, and dances | Group | 12 | 2/week (3 + 3) | 60 min. | NS |

CAREGIVERS' TRAINING

| | | | | | | | | | | |
|---|---|---|---|---|---|---|---|---|---|---|
| Götell et al. | 2003 | Sweden | 9/severe & 5 caregivers | -Non-verbal aspects of communication: posture, movement and sensory awareness | -Background music -Caregiver singing | Individual | 3 | NS | 6–22 min. | RN |
| Gallagher | 2011 | USA | 24 hospice staff: 12 RN, 5 CNA, 5 SW, 2 Chaplain | To teach to implement IM to reduce agitation | IM—singing/listening | Group | 1 | | 60 min. | Hospice staff |
| Hammar et al. | 2011 | Sweden | 10 pairs PWD at the severe stage and caregiver | -Verbal and nonverbal communication: eye contact during the caring activity 'getting dressed' | -MTC: singing songs from early years of PWD accompanied by body movements | Group | Daily | 2 months | NS | CNA |
| Hsu et al. | 2015 | UK | 17/ Moderate-severe | -BPSD symptoms | Active MT: singing, improvisation, talking | Individual | 9-22 | 1/week (5 months) | 30 min. | MTx |

BPSD: Behavioral and Psychological Symptoms of Dementia; CNA: Certified Nurse Assistant; IM: Individualized Music; MT: Music Therapy; MTC: Music Therapeutic Caregiving; MTx: Music Therapist; NA: Nurses aides; NS: Not specified; OT: Occupational Therapist; QoL: Quality of Life; RN: Registered Nurses; RSS: Rhythmic Sensory Stimulation; SW: Social Worker.

[1] Results are reported only for the participants with GDS 6–7.

Sixteen of the 25 included studies were MT studies where the interventions were led by professional music therapists. Thirteen of the studies included group interventions, 10 focused on individual interventions and two did not specify the type of intervention. Table 12.1 lists all the studies and their key features.

## Symptoms of People in Late Stages of Dementia and Areas of Intervention in Music Therapy

The needs of people in late-stage dementia are identified according to the domains addressed in the various music interventions specified in the literature.

### Physiological Domain

Studies focusing on the impact of MT on different physiological parameters for people with dementia are scarce in the literature. In fact, only one study which addressed physiological outcomes for people in late stage dementia was identified in the literature search (Takahashi & Matsushita, 2006). This study investigated the lasting effects of MT on cortisol level in saliva and blood pressure, as indicators of stress in people in the moderate and late stages of the dementia. Results showed that increased salivary cortisol levels immediately after MT seemed to be related to a higher level of participation and higher intelligence scores as measured by the Revised Hasegawa Dementia Scale (HDS-R). In addition, systolic blood pressure was significantly lower in people who received MT after two years of treatment. Although systolic blood pressure has the tendency to increase with age, the fact that MT may have a lowering effect on systolic blood pressure over time could imply that participation in MT may help prevent cardiac and cerebral diseases.

### Cognitive Domain

Due to the degenerative nature of dementia, impairments in memory, thinking, and behavior continue to decline as the disease progresses.

Moreover, people with late-stage dementia often lose their ability to speak or communicate. The cognitive areas addressed in the studies reviewed include overall cognitive function, alert behavior, communication/expressive skills (specifically the ability to maintain a conversation) and short-term memory.

Regarding the effectiveness of MT interventions on overall cognitive function, Clements-Cortes and colleagues (2016) evaluated the effects of rhythmic sensory stimulation on cognition of people with AD at mild, moderate, and late stages of the disease. This technique is based on the use of the sound-driven vibrotactile stimulation of the somatosensory system. An increase in alertness, awareness of surroundings, interaction, and stimulation of discussion/storytelling were observed in the majority of participants post intervention. Daily music-based seated dance sessions for 3 months also led to positive results on cognition and behavior in women with dementia (Van de Winckel, Feys, De Weerdt, & Dom, 2004). This music intervention led to significantly higher scores on the Mini-Mental State Exam, specifically in the fluency test, in patients with dementia. However, no behavior changes were found. The authors suggested that the improvement in the fluency test in this study could be explained as an increase in information processing speed, making the latent information available faster to the patient.

Some studies compared music interventions or MT with other types of treatment such as recreational therapy and multisensory stimulation environment on the cognitive skills of people with late dementia, such as expressive and communicative skills (van Bruggen-Rufi, Vink, Wolterbeek, Achterberg, & Roos, 2017) and cognitive status, agitation, and emotional status (Sánchez et al., 2016). In both studies MT and music intervention had no additional beneficial effects on improving any of the cognitive skills or on reducing behavioral problems in comparison to other types of therapy.

The ability to sing music that was learned in youth often remains intact throughout the progression of dementia. People with dementia often use nonverbal communication for self-expression and interaction as a result of the decline in speech production and comprehension. Dassa and Amir (2014) evaluated the role of singing familiar songs from the participants' past, related to their social and national identity. The results of their study showed that this intervention encouraged conversation among people with middle- to late-stage AD and elicited rich memories. Furthermore, it was observed that conversation related to the singing was extensive, and group singing encouraged spontaneous responses such as positive feelings and a sense of accomplishment and belonging. The combination of music with touch during MT has also been evaluated to enhance communication, especially nonverbal communication, among people in all stages of dementia (Belgrave, 2009). Expressive and instrumental touch, used to assist in the completion of a musical task and to foster alert behavior of the participants, were compared. Expressive touch, defined as a nurturing or caring touch applied to the shoulder, arm, or hand was significantly more effective in eliciting and maintaining alert behavior states. The results of the studies in this area indicate the effectiveness of MT, especially to maintain alertness and to foster communication and interaction among group members and between participants and music therapists.

## Social/Emotional Domain

Engagement in a wide range of musical activities continues to be rated highly among the various events enjoyed by residents in most long-term care facilities. The reported evidence strongly suggests that the ability to enjoy and engage in musical elements in a meaningful way remains strong even after other cognitive abilities have practically disappeared (Gold, 2014; Shibazaki & Marshall, 2017).

People with late-stage dementia often experience a decline in motivation, thereby showing less initiative to participate in activities (Clair, 1996). There is a need for sensory stimulation and interaction with the environment in order to combat this decline, to maintain skills within these areas, and to improve the quality of life of these individuals. But it is important to assess participants' abilities and to know the most appropriate types of music stimulation for people with dementia in order to provide an optimal sensory environment and foster participation for people in the late stages of dementia. Choral singing is an activity that has shown benefits among older adults in several areas: psychophysiological, social, and emotional as well as for persons with dementia and their caregivers (Clements-Cortés, 2015a, 2015b). Moreover, there is an emerging body of evidence that suggests that music which is specifically customized to the preferences of the individual listener with dementia can have a positive impact (Gerdner, 2005, 2012; Gerdner & Schoenfelder, 2010).

Another aspect that is of particular interest when considering providing music and MT based activities is the type of music intervention: active or receptive. Active music interventions seem to be more effective than, or as effective as, receptive music conditions in promoting positive participation (Raglio et al., 2015; Sakamoto et al., 2013). However, participants with more advanced dementia appear to be more alert during active music interventions (Lancioni et al., 2015). Similar results were observed in a dementia study by Groene (2001) who also evaluated different accompaniment styles (simple vs. complex) besides the effects of live versus recorded music presentations on attention and responsive behaviors during group singing. Participants attended and stayed longer in the sessions, read lyrics more, and gave more compliments and applause with live music and complex accompaniments. However, these different modes of presenting music did not make a difference to the amount of singing by participants. Along the same lines, Cevasco and Grant (2006) studied the value of instruments or combination of instruments (djembe, keyboard, autoharp, and guitar) used by the therapist to elicit responses during singing, movement, and playing rhythm instruments with elderly persons in middle to later stages of dementia. The greatest participation occurred during rhythm-based activities. Moreover, there was an increase in participation during movement and rhythm activities when sensory stimulation was less (i.e., singing without instrumental accompaniment). When there was an overload of sensory stimulation, participation decreased. In this study, a cappella singing and then singing with djembe resulted in greatest overall participation. During movement activities the greatest amount of movement, either movement alone or singing and movement combined, occurred during a cappella singing and secondly during singing with djembe accompaniment. These results suggest that music presentation should be simple for people in late-stage dementias in order to maximize participation and minimize the potential for sensory overload.

Engagement is crucial to fostering social interaction and communication. Communication is essential for human beings because it facilitates the sharing of information and helps in developing relationships. For people with dementia who may also suffer from aphasia, nonverbal communication may play an important role in sharing experiences, and music is powerful when it comes to facilitating recognizable forms of communication and to engage in mutual interaction. Ridder and Gummesen (2015) explored an improvisation method known as "extemporizing" with a person with moderate-severe dementia and aphasia to understand how music therapy may facilitate communication and dialogue. They concluded that musical improvisation may enhance free expression, and the improvisation technique of extemporization may be specifically important in people with moderate-late dementia and aphasia, as it takes its starting point from preferred musical material and well-known musical forms. Individualized MT also showed improvements in communication, expression of positive emotions, and situational well-being (Schall, Haberstroh, & Pantel, 2015). The findings of the studies included in this section seem to indicate that overall interactive music intervention may be effective in enhancing communication and social connection between people with slate dementia and other people, leading to improved quality of life among elderly individuals with late dementia.

## Behavioral-Psychological Symptoms of Dementia

It is estimated that the prevalence of behavioral and psychological symptoms of dementia (BPSD) ranges from 25 to 80% (Sadak, Katon, Beck, Cochrane, & Borson, 2014) and these symptoms increase as the severity of dementia progresses. The majority of people with AD in the later stages of the disease show signs of psychiatric distress and a range of aberrant behavioral patterns such as agitation, aggression, repetitive vocalizations, yelling, and psychosis (International Psychogeriatric Association, 2011). Several studies and systematic reviews of nonpharmacological interventions to treat behavioural disturbances in older patients with dementia suggest that MT is a promising nonpharmacological approach to ameliorate these symptoms (Gold, 2014; Gómez-Romero et al., 2017; Ray & Mittleman, 2015). Moreover, the use of prerecorded music outside of formal MT settings has become a very popular intervention in health-care contexts for people living with dementia. Although results reveal that prerecorded music, facilitated by nonmusic therapists, can be effective in reducing a variety of affective and behavioral symptoms, in particular agitation, they need to be interpreted with caution since more empirical studies are needed in order to clearly establish the efficacy of this type of intervention (Garrido et al., 2017). The BPSD addressed in the studies reviewed include agitation, apathy and social withdrawal.

**Agitation.** Several studies have been conducted focusing on the potential benefits of music-based interventions and MT to decrease agitation. Gerdner (2012) explains an evidence-based individualized music protocol to treat anxiety and agitation for people with late dementia. She summarizes the results of different studies conducted around the world (Taiwan, China, France, & USA) that support this intervention which involves listening to individualized music based on the patient's specific music preferences. In addition, results suggest that this intervention can facilitate meaningful interaction between people with dementia and others (Gerdner, 2005).

On the other hand, Tuet and Lam (2006) evaluated the effectiveness of 3 weeks of active group MT on agitation in people in the moderate-late stage of dementia. Results showed a significant reduction in Neuropsychiatric Inventory and Cohen-Mansfield inventory scores immediately after MT although the effects did not last after the termination of MT treatment. Similar results were obtained by Sakamoto, Ando, and Tsutou (2013) when comparing active versus passive music interventions on the stress level and BPSD of patients in the late stage of dementia. Interactive music interventions showed the strongest beneficial effects, suggesting that this type of intervention can restore residual cognitive and emotional function, increase social engagement for people with late dementia, and improve quality of life (Sakamoto et al., 2013). Another study in Iceland also showed the efficacy of using live music to treat disruptiveness, aggression, and anxiety in people with late-stage dementia (Svansdottir & Snaedal, 2006). However, the effects of this intervention lasted for 4 weeks after the intervention. Similar results were obtained for an active MT group in a study conducted by Raglio et al. (2008). The decrease in agitation also persisted one month post MT intervention. In addition, an increase in smiles, body movement, and singing were observed, suggesting improvements in communication between patient and music therapist.

In a later study, Raglio et al. (2010) assessed the efficacy of active MT treatment based on three working cycles of 1 month spaced out by 1 month of no treatment on the behavioural disturbances of people with severe dementia. Delusions, agitation, and apathy significantly improved only in the MT group, while depression, anxiety and irritability improved significantly both in the experimental and in the control groups who received standard care (educational and entertainment activities, such as reading a newspaper, performing physical activities, etc.). Results also showed maintenance of improvements in the MT group 2 months after the treatment. When comparing the effects of individualized active MT and music listening on BPSD of dementia, Raglio et al. (2015) found that the group that attended individualized active MT showed the greatest improvements in the Neuropsychiatric Inventory scores. Moreover, a decrease in stress and an increase in positive emotional responses were some of the short-term effects of interactive participation in an activity using individualized music associated with special memories

(Sakamoto et al., 2013). These changes were observed 2 weeks from the end of the intervention but had disappeared by 3 weeks after the intervention period, indicating that the intervention should be regularly conducted to exert continued beneficial effects.

The level of agitation and disruptiveness for people with dementia also decreased significantly after 6 weeks of individual MT intervention (Ridder, Stige, Qvale, & Gold, 2013). Moreover, the prescriptions of psychotropic medication did not increase. According to these results, MT seems to increase engagement and engagement duration, specifically in one-on-one interventions. These results are supported in a review by Millán-Calenti et al. (2016) which emphasizes the idea that MT is optimal for the management of agitation in institutionalized patients with moderately severe and late AD, particularly when the intervention includes individualized and interactive music.

Ray and Mittleman (2015) evaluated depressive symptoms, agitation, and wandering in a group of residents with moderate-late dementia after 2 weeks of MT intervention. Results determined that symptoms of depression and agitation were significantly reduced, but there was no change for wandering. Other studies addressing BPSD have compared music interventions, implemented by nonmusic therapists, with other types of interventions: cooking (Narme et al., 2014), family presence (Garland, Beer, Eppingstall, & O'Connor, 2007; Conner & Daniel, 2007), multisensory stimulation environment (mentioned in the cognitive domain section; Sánchez et al., 2016) on various emotional, cognitive, and behavioral measures. The results of these studies showed similar positive effects for the various interventions.

With the advancement of technology, multimedia devices such as the "Memory Box," which offers digitalized music, photographs, movies, and messages chosen or made by family members to reflect residents' backgrounds and preferences, have been introduced in dementia care. Their effectiveness in improving some of the BPSD symptoms, such as agitation, have been evaluated (Davison et al., 2016). Significant reductions were observed in depressive and anxiety symptoms during the course of the intervention although more severely impaired residents needed help to use the device.

Although passive music-based interventions and active MT are used in the treatment of BPSD of people in the late stages of the dementia, the results of the studies reviewed suggest that interactive MT is more effective than receptive interventions to reduce some symptoms, such as agitation. In addition, the positive effects of active MT seem to last for longer periods of time than those of passive music interventions.

**Apathy and social withdrawal.** The cognitive deficits which accompany the progression of dementia, including language deterioration, often lead to losses in social relationships, and consequently social withdrawal and apathy. Holmes, Knights, Dean, Hodkinson, and Hopkins (2006) conducted a study which showed an improvement in positive engagement of people in the moderate or late stages of the disease as a result of their participation in live interactive music. In addition, the results also show that the benefits of prerecorded music, in terms of the positive engagement, would appear to be of limited value, particularly when the participants have late dementia.

## Quality of Life

Due to a reduced threshold for handling stressful environmental influences as well as BPSD, patients with progressive cognitive impairment typically experience poor quality of life. Quality of life of elderly persons in residential care has been researched and characterized into four dimensions: participation, mental well-being, physical well-being, and residential conditions. Van der Vleuten, Visser, and Meeuwesen (2012) evaluated the effects of intimate live music performances on the quality of life dimensions of participation and mental well-being for people with dementia. The performances seemed to have a positive effect on participation for people with mild and late dementia. They also had a positive impact in the mental well-being on the group of people with mild dementia, although significant changes were observed in the affect of those with late dementia. In addition, the participants enjoyed the intimate live music performance and wanted to see it again. The effects of group participation in MT on QoL, participation, and affect were also evaluated by Sole, Mercadal-Brotons, Galati, and De Castro (2014). Although participation was high throughout the program, QoL, as measured by a researcher-designed questionnaire, did not improve from pre to post treatment, but participants gained improvements in affect.

The use of technology is also part of nonpharmacological interventions for people with dementia to foster quality of life. Thus, the effects of video-MT (folk music associated with the projection of local festival movies) was analyzed on the perceived quality of life in patients with AD in various stages of the disease (Rubbi et al., 2016). Patients with severe cognitive impairment did not gain any improvement in QoL whereas all other patients showed a good response to this intervention.

## Training Caregivers to Deliver Music-Based Interventions

People in the late stages of dementia may present challenges to their caregivers who continue to provide for them at home for as long as possible. There is a growing recognition for the need to include family members in the planning and implementation of care for their family member with dementia. Furthermore, family members usually remain involved in their loved ones' care and they also have the need to continue a personal relationship with them even when care can no longer be provided at home.

Music therapists have become progressively more involved in training professional caregivers in the use of music-based interventions, such as unaccompanied singing and making individualized playlists with preferred music. These trainings empower professional caregivers to use music more confidently (Gallagher, 2011), manage agitation (Gerdner, 2005), facilitate interaction and communication between caregivers and people with dementia, increase patient cooperation during caregiving activities, and increase meaningful interactions between patients and caregivers (Götell, Brown, & Ekman, 2002; Hammar, Emami, Engström, & Götell, 2011; Hsu et al., 2015; Ray & Fitzsimmons, 2014).

## Methodological Issues

The dosage of MT (number and frequency of sessions) varied considerably between studies. In the studies reviewed, the frequency of sessions ranged from 1 to 3 times a week, with the total number of sessions ranging between 20 and 32 per program evaluated. However, the duration of the MT sessions was quite similar across studies, commonly 30–45 minutes. The results of the studies included in this chapter also show that persons in late-stage dementia become more responsive to the intervention when they are involved in MT over time. The more MT offered, the better and more sustainable the short-term and long-term outcomes, although changes are more difficult to maintain as the dementia becomes more severe (Ridder, Stige, Qvale, & Gold, 2013; Sakamoto et al., 2013).

Some consistencies in methodology observed in the studies related to positive outcomes, include

1. The use of patient-preferred music for active MT and for music listening interventions (personalized or individualized music programs).
2. Live music interventions, which involve patient-therapist interaction, were more effective in reducing agitation than passive listening to prerecorded music. However, both passive and interactive music interventions were equally effective in reducing stress and inducing relaxation in individuals with late dementia.
3. Study settings included individual and small groups; however, the accumulation of clinical and empirical data provides a growing body of evidence that supports the use of individualized music (in music listening as well as in live music) for persons with late-stage dementia.
4. Areas most often addressed in studies with people with late dementia included engagement and participation, interaction and communication (including caregivers), management and reduction of BPSD (aggression, agitation, anxiety, disruptiveness when being involved in care routines such as bathing), and quality of life.

According to this review, the music-based interventions that showed positive results were:

- Rhythmic brain stimulation to improve cognitive functioning (Clements-Cortés et al., 2016).
- Personalized playlists to reduce agitation and aggressiveness at bath time and while performing other activities of daily living (Gerdener, 2012). The active MT techniques which were most appropriate to the physical and cognitive abilities of people in late-stage dementia include:
  - Rhythm-based activities and playing instruments to maintain engagement and participation (Cevasco & Grant, 2006; Holmes, Knights, Dean, Hodkinson, & Hopkins (2006), as well as
  - Singing (with and without accompaniment) to foster interaction with others (Groene, 2001; Ridder & Gummesen, 2015). Singing is accessible to all persons, regardless of musical background and training (e.g. caregivers). Caregiver singing may also lead to decreases in symptoms of agitation and depression.

Other nonmusical methodological considerations mentioned in some of the studies reviewed include therapist's presence and consistency of sessions. The importance of the therapist maintaining an attentive presence throughout the session—sitting close to the person with dementia and making direct eye contact—has been emphasized to maintain a strong connection with the person and increase alertness and engagement (Tomaino, 2013). Expressive and instrumental touch have been used during MT sessions to increase rapport and communication between carer and music therapist while fostering alert responses of persons in late-stage dementia (Belgrave, 2009).

## Summary and Conclusions

Although all functional areas have been addressed in the various studies reviewed in this chapter, many studies focused on evaluating diverse music interventions to improve BPSD, which are perhaps the most challenging feature of dementia, especially in the late stage. It is important to note that studies included have been conducted in very diverse countries with people with dementia from different cultural backgrounds, and certain music protocols seem to be equally effective to address specific issues (e.g., agitation) despite geographic and cultural differences.

People with a dementia diagnosis inevitably come to the end of their lives with severe physical, emotional, and social debilitation. These people lose the physical capacities to ambulate and to perform purposeful physical, social, and cognitive activities. They are, out of necessity, confined either to bed or reclining chairs, and their lack of mobility brings on other chronic and acute health impairments. Most people with late dementia inevitably require placement in institutional settings where their nursing care needs can be met (Clair, 1996). These facilities therefore need to provide programs which meet quality of life needs in total health care. Music therapy and music activities can be effective nonpharmacological interventions to address these needs of these persons, but what is clear from the studies reviewed is that the use of music needs to be adapted to each person's characteristics. Therefore, individual needs must be considered (Ray & Mittleman, 2015).

From a person-centered or need-driven, dementia-compromised behavior model perspective, the needs of participants may be met by the music therapists' acknowledgement of each individual's background by singing familiar songs and reminiscing about family relations, life experiences, or childhood memories (Penrod et al., 2007). The use of MT for BPSD has been widely studied in observational trials for decades (Tsoi, Chan, Ng, Lee, Kwok, & Wong, 2018). As a nonpharmacological treatment, MT is recommended as an intervention to reduce behavioral symptoms (Livingston et al., 2014; Ueda et al., 2013), though there is a need to document the procedures undertaken for music-based interventions. Fortunately, more current studies evaluate specific music-based protocols aiming to improve specific symptoms in a variety of domains, as this chapter reflects. The scientific literature is still heterogeneous in terms of evidence-based music interventions and who implements them: professional music therapists, other professionals, and caregivers. Raglio and colleagues (2014) propose the Global Music Approach to persons with Dementia (GMA-D) as a structured intervention model which combines a whole variety of music-based intervention strategies. This model requires the presence of an adequately trained music therapist who, in collaboration with the involved clinical staff and caregivers, guarantees the correct planning and management of the music interventions and their evaluation. A well-coordinated approach in the use of music interventions can lead to better results in reducing behavioral disturbances, stimulating cognitive functions, and increasing the overall quality of life of people in late-stage dementia.

# References

Abraha, I., Rimland, J. M., Trotta, F. M., Aquila, G. D., Cruz-jentoft, A., Petrovic, M., . . . Cherubini, A. (2017). Systematic review of systematic reviews of nonpharmacological interventions to treat behavioural disturbances in older patients with dementia. *The SENATOR-OnTop series.* https://doi.org/10 1136/bmjopen-2016-012759

Alzheimer's Disease International & World Health Organization. (2017). *Dementia: A public health priority.* Retrieved from http://apps.who.int/iris/bitstream/10665/75263/1/9789241564458_eng.pdf?ua=1https://www.alz.org/documents_custom/2017-facts-and-figures.pdf

Alzheimer's Society. (2018). The later stages of dementia. Retrieved from https://www.alzheimers.org.uk/info/20073/how_dementia_progresses/103/the_later_stages_of_dementia/3

American Psychiatric Association. (2013). *Diagnostic and statistical manual of mental disorders* (5th ed.). Lake St. Louis, MO: American Psychiatric Association.

Ballard, C., Khan, Z., Clack, H., & Corbett, A. (2011). Nonpharmacological treatment of Alzheimer disease. *Canadian Journal of Psychiatry, 56,* 589–595.

Belgrave, M. (2009). The effect of expressive and instrumental touch on the behavior states of older adults with late-stage dementia of the Alzheimer's type and on music therapist's perceived rapport. *Journal of Music Therapy, 46*(2), 132–146.

Boso, M., Politi, P., Barale, F., & Enzo, E. (2006). Neurophysiology and neurobiology of the musical experience. *Functional Neurology, 21*(4), 187–191.

Cevasco, A., & Grant, R. (2006). Value of musical instruments used by the therapist to elicit responses from individuals in various stages of Alzheimer's disease. *Journal of Music Therapy, 43,* 226–246.

Clair, A. A. (1996). The effect of singing on alert responses in persons with late stage dementia. *Journal of Music Therapy, 33*(4), 234–247.

Clair, A. A., & Ebberts, G. (1997). The effects of music therapy on interactions between caregivers and their care receivers with late stage dementia. *Journal of Music Therapy, 34*(3), 148–164.

Clements-Cortés, A. (2015a). Clinical effects of choral singing for older adults. *Music and Medicine, 7*(4), 7–12.

Clements-Cortés, A. (2015b). Singing for health, connection and care. *Music and Medicine, 7*(4), 13–23.

Clements-Cortés, A., Ahonen, H., Evans, M., Freedman, M., & Bartel, L. (2016). Short-term effects of rhythmic sensory stimulation in Alzheimer's disease: An exploratory pilot Study. *Journal of Alzheimer's Disease, 52,* 651–660. doi:10.3233/JAD-160081.

Connor, O., & Daniel, W. (2007). A comparison of two treatments of agitated behavior in nursing home residents with dementia: Simulated family presence and preferred music. *The American Journal of Geriatric Psychiatry, 15*(6), 514–521.

Cooper, C. (2014). A systematic review of the clinical effectiveness and cost-effectiveness of sensory, psychological and behavioural interventions for managing agitation in older adults with dementia. *Health Technology Assessment (Winchester, England), 18*(39), 1–226. https://doi.org/10.3310/hta18390

Dassa, A., & Amir, D. (2014). The role of singing familiar songs in encouraging conversation among people with middle to late stage Alzheimer's disease. *Journal of Music Therapy, 51*(2), 131–153. https://doi.org/10.1093/jmt/thu007

Davison, T. E., Nayer, K., Coxon, S., de Bono, A., Eppingstall, B., Jeon, Y. H., ... O'Connor, D. W. (2016). A personalized multimedia device to treat agitated behavior and improve mood in people with dementia: A pilot study. *Geriatric Nursing, 37*(1), 25–29. https://doi.org/10.1016/j.gerinurse.2015.08.013

Dyer, S. M., Harrison, S. L., Laver, K., Whitehead, C., & Crotty, M. (2018). An overview of systematic reviews of pharmacological and non-pharmacological interventions for the treatment of behavioral and psychological symptoms of dementia. *International Psychogeriatrics, 30*(3), 295–309.

Engeström, Y., Miettinen, R., & Punamäki, R. (Eds.). (1999). *Perspectives on activity theory.* Cambridge, UK: Cambridge University Press.

Fisher Center for Alzheimer's Research Foundation. (2017). Retrieved from https://www.alzinfo.org/

Gallagher, M. (2011). Evaluating a protocol to train hospice staff in administering individualized music, *17*(4), 195–202.

Garland, K., Beer, E., Eppingstall, B., & O'Connor, D. W. (2007). A comparison of two treatments of agitated behavior in nursing home residents with dementia: Simulated family presence and preferred music. *The American Journal of Geriatric Psychiatry, 15*(6), 514–521.

Garrido, S., Dunne, L., Chang, E., Perz, J., Stevens, C., & Haertsch, M. (2017). The use of music playlists for people with dementia: A critical synthesis. *Journal of Alzheimer's Disease, 60,* 1129–1142.

Gerdner, L. A. (2005). Use of individualized music by trained staff and family: Translating research into practice. *Journal of Gerontological Nursing, 31*(6), 22–30.

Gerdner, L. A. (2012). Individualized music for dementia: Evolution and application of evidence-based protocol. *World Journal of Psychiatry, 2*(2), 26–32. https://doi.org/ 10.5498/wjp.v2.i2.26

Gerdner, L. A., & Schoenfelder, D. P. (2010). Evidence-based guide-lines: Individualized music for elders with dementia. *Journal of Gerontological Nursing, 36*(6), 7–15.

Gold, K. (2014). But does it do it any good? Measuring the impact of music therapy on people with advanced dementia: (Innovative practice). *Dementia, 13,* 258–264.

Gómez-Romero, M., Jiménez-Palomares, M., Rodríguez-Mansilla, J., Flores-Nieto, A., Garrido-Ardila, E. M., & Gonzealez-LópezArza, M. V. (2017). Beneficios de la musicoterapia en las alteraciones conductuales de la demencia. Revisión sistemática. *Neurología, 32*(4), 253–263.

Götell, E., Brown, S., & Ekman S. L. (2002). Caregiver singing and background music in dementia care. *Western Journal of Nursing Research, 24*(2), 195–216.

Groene, R. (2001). The effect of presentation and accompaniment styles on attentional and responsive behaviors of participants with dementia diagnoses. *Journal of Music Therapy, 38*(1), 36–50.

Gross, S. M., Danilova, D., Vandehey, M. A., & Diekhoff, G. M. (2015). Creativity and dementia: Does artistic activity affect well-being beyond the art class? *Dementia, 14*(1), 27–46. https://doi.org/10.1177/1471301213488899

Hammar, L. M., Götell, E., & Engström, G. (2011). The impact of caregivers' singing on expressions of emotion and resistance during morning care situations in persons with dementia: an intervention in dementia care. *Journal of Clinical Nursing, 20*(7–8), 969–978. doi:10.1111/j.1365-2702.2010.03386

Holmes, C., Knights, A., Dean, C., Hodkinson, S., & Hopkins, V. (2006). Keep music live: Music and the alleviation of apathy in dementia subjects. *International Psychogeriatrics, 18*(4), 623–630. https://doi.org/10.1017/S1041610206003887

Hsu, M. H., Flowerdew, R., Parker, M., Fachner, J., & Odell-Miller, H. (2015) Individual music therapy for managing neuropsychiatric symptoms for people with dementia and their carers: a cluster randomised controlled feasibility study. *BMC Geriatrics, 15,* 84. doi: 10.1186/s12877-015-0082-4.

International Psychogeriatric Association. (2011). *IPA Complete guides to behavioral and psychological symptoms of dementia (BPSD).* Milwaukee, WI: International Psychogeriatric Association.

Jacobsen, J. H., Stelzer, J., Fritz, T. H., Che, G., Joie, R. La, & Turner, R. (2015). Why musical memory can be preserved in advanced Alzheimer's disease. *Brain 138*(8), 2438–2450. https://doi.org/10.1093/awv148

Janata, P. (2009). The neural architecture of music-evoked autobiographical memories. *Cerebral Cortex, 19*(11), 2579–2594. https://doi.org/10.1093/cercor/bhp008

Karli, B. E., Young, D., & Dash, K. (2017). Empowering the dementia care workforce to manage behavioral symptoms of dementia: Development and training outcomes from the VOICE dementia care program. *Gerontology and Geriatrics Education, 38*(4), 375–391.

Lancioni, G. E., Bosco, A., De Caro, M. F., Singh, N. N., O'reilly, M. F., Green, V. A., . . . Zonno, N. (2015). Effects of response-related music stimulation versus general music stimulation on positive participation of patients with Alzheimer's disease. *Developmental Neurorehabilitation, 18*(3), 169–176. https://doi.org/10.3109/17518423.2013.802388.

Lin, P. H., Chen, H. H., Chen, N. C., Chang, W. N., Huang, C. W., Chang, Y. T., . . . Chang, C. C. (2016). Anatomical correlates of non-verbal perception in dementia patients. *Frontiers in Aging Neuroscience, 8,* 207.

Livingston, G., Kelly, L., Lewis-Holmes, E., Baio, G., Morris, S., Patel, N., & Cooper, C. (2014). Non-pharmacological interventions for agitation in dementia. Systematic review of randomized controlled trials. *The British Journal of Psychiatry, 205*(6), 436–442.

Millan-Calenti, J. C., Lorenzo-López, L., Alonso-Búa, B., de Labra, C., González-Abraldes, I., & Maseda, A. (2016). Optimal non-pharmacological management of agitation in Alzheimer's disease: Challenges and solutions. *Clinical Interventions in Aging, 11,* 175–184.

Narme, P., Clément, S., Ehrlé, N., Schiaratura, L., Vachez, S., Courtaigne, B., . . . Samson, S. (2014). Efficacy of musical interventions in dementia: Evidence from a randomized controlled trial. *Journal of Alzheimer's Disease, 38*(2), 359–369. https://doi.org/10.3233/JAD-130893.

Olley, R., & Morales, A. (2017). Systematic review of evidence underpinning non-pharmacological therapies in dementia. *Australian Health Review 42*(4), 361–369. doi:10.1071/AH16212

Penrod, J., Yu, F., Kolanowski, A., Fick, D. M., Loeb, S. J., & Hupcey, J. E. (2007). Reframing person-centered nursing care for persons with dementia. *Research and Theory for Nursing Practice, 21*(1), 57–72.

Phinney, A., Chaudhury, H., & O'Connor, D. L. (2007). Doing as much as I can do: The meaning of activity for people with dementia. *Aging & Mental Health, 11*(4), 384–393.

Prince, M., Ali, G. C., Guerchet, M., Prina, A. M., Albanese, E., & Wu. Y. T. (2016). Recent global trends in the prevalence and incidence of dementia, and survival with dementia. *Alzheimer's Research & Therapy, 8,* 23.

Prince, M., Wimo, A., Guerchet, M., Ali, G. C., Wu, Y. T., & Prima, M. (2015). World Alzheimer Report 2015. The global impact of dementia an analysis of prevalence, incidence, cost and trends. Retrieved from https: //www.alz.co.uk/research/WorldAlzheimerReport2015. pdf

Raglio, A., Bellandi, D., Baiardi, P., Gianotti, M., Ubezio, M. C., Zanacchi, E., . . . Stramba-Badiale, M. (2015). Effect of active music therapy and individualized listening to music on dementia: A multicenter randomized controlled trial. *Journal of the American Geriatrics Society, 63*(8), 1534–1539. https://doi.org/10.1111/jgs.13558

Raglio, A., Bellelli, G., Traficante, D., Gianotti, M., Ubezio, M. C., Gentile, S., . . . Manual, S. (2010). Efficacy of music therapy treatment based on cycles of sessions: A randomised controlled trial. *Aging and Mental Health, 14*(8), 900–904. https://doi.org/10.1080/13607861003713158

Raglio, A., Bellelli, G., Traficante, D., Gianotti, M., Ubezio, M. C., Villani, D., & Trabucchi, M. (2008). Efficacy of music therapy in the treatment of behavioral and psychiatric symptoms of dementia. *Alzheimer Disease and Associated Disorders, 22*(2), 158–162. https://doi.org/10.1097/WAD.0b013e3181630b6f

Raglio, A., Filippi, S., Bellandi, D., & Stramba-Badiale, M. (2014). Global music approach to persons with dementia: Evidence and practice. *Clinical Interventions in Aging, 9,* 1669–1676. https://doi.org/10.2147/CIA.S71388

Ray, K., & Fitzsimmons, S. (2014). Music-assisted bathing: Making shower time easier for people with dementia. *Journal of Gerontological Nursing, 40*(2), 9–13. doi:10.3928/00989134-20131220-09

Ray, K. D. & Mittelman, M. S. (2015). Music therapy: A nonpharmacological approach to the care of agitation and depressive symptoms for nursing home residents with dementia. *Dementia, 16*(6), 689–710. https://doi.org/10.1177/1471301215613779

Ridder, H. M. O, & Gummensen, E. (2015). The use of extemporizing in music therapy to facilitate communication in a person with dementia: An explorative case study. *Australian Journal of Music Therapy, 26*, 1–25.

Ridder, H. M. O., Stige, B., Qvale, L. G., & Gold, C. (2013). Individual music therapy for agitation in dementia: An exploratory randomized controlled trial. *Aging & Mental Health, 17*(6), 667–678. https://doi.org/10.1080/13607863.2013.790926

Rubbi, I., Magnani, D., Naldoni, G., Di Lorenzo, R., Cremonini, V., Capucci, P., . . . Ferri, P. (2016). Efficacy of video-music therapy on quality of life improvement in a group of patients with Alzheimer's disease: A pre-post study. *Acta Bio-Medica: Atenei Parmensis, 87*(4–S), 30–37.

Sadak, T. I., Katon, J., Beck, C., Cochrane, B. B., & Borson, S. (2014). Key neuropsychiatric symptoms in common dementias: Prevalence and implications for caregivers, clinicians, and health systems. *Research in Gerontological Nursing, 7*(1), 44–52. doi:10.3928/19404921-20130918-01

Sakamoto, M., Ando, H., & Tsutou, A. (2013). Comparing the effects of different individualized music interventions for elderly individuals with severe dementia. *International Psychogeriatrics, 25*(5), 775–784. https://doi.org/10.1017/S1041610212002256

Sánchez, A., Maseda, A., Marante-Moar, M. P., de Labra, C., Lorenzo-López, L., & Millán-Calenti, J. C. (2016). Comparing the effects of multisensory stimulation and individualized music sessions on elderly people with severe dementia: A randomized controlled trial. *Journal of Alzheimer's Disease, 52*(1), 303–315.

Schall, A., Haberstroh, J., & Pantel, J. (2015). Time series analysis of individual music therapy in dementia. *GeroPsych, 28*(3), 113–122. https //doi.org/10.1024/1662-9647/a000123

Shibazaki, K., & Marshall, N. A. (2017). Exploring the impact of music concerts in promoting well-being in dementia care. *Aging and Mental Health, 21*(5), 468–476. https://doi.org/10.1080/13607863.2015.1114589

Sjogren, K., Lindkvist, M., Sandman, P. O., Zingmark, K., & Edvardsson, D. (2013). Person- centredness and its association with resident well-being in dementia care units. *Journal of Advanced Nursing, 69*, 2196–2205. doi:10.1111/jan.12085; 10.1111/jan.12085

Sole, C., Mercadal-Brotons, M., Galati, A., & De Castro, M. (2014). Effects of group music therapy on quality of life, affect, and participation in people with varying levels of dementia. *Journal of Music Therapy, 51*(1), 103–125. https://doi.org/10.1093/jmt/thu003

Svansdottir, H. B., & Snaedal, J. (2006). Music therapy in moderate and severe dementia of Alzheimer's type: A case-control study. *International Psychogeriatrics, 18*(4), 613–621.

Takahashi, T., & Matsushita, H. (2006). Long-term effects of music therapy on elderly with moderate/severe dementia. *Journal of Music Therapy, 43*(4), 317–333.

Tomaino, C. M. (2013). Meeting the complex needs of individuals with dementia through music therapy. *Music and Medicine, 5*(4), 234–241 https://doi.org/10.1177/1943862113505775.

Tsoi, K., Chan, J., Ng, Y., Lee, M., Kwok, T., & Wong, S. (2018). Receptive music therapy is more effective than interactive music therapy to relieve behavioral and psychological symptoms of dementia: A systematic review and meta-analysis. *Journal of the American Medical Directors Association, 19*(7), 568–576.

Tuet, R. W., & Lam, L. C. (2006). A preliminary study of the effects of music therapy on agiation in Chinese patients with dementia. *Hong Kong Journal of Psychiatry, 16*, 87–91.

Ueda, T., Suzukamo, Y., Sato, M., & Izumi, S. (2013). Effects of music therapy on behavioral and psychological symptoms of dementia: A review and meta-analysis. *Ageing Research Reviews, 12*(2), 628–641. doi:10.1016/j.arr.2013.02.003

Van Bruggen-Ruffi, M., Vink, A., Wolterbeek, R., Achterberg, W., & Roos, R. (2017). The effect of music therapy in patients with Huntington's disease. A randomized controlled trial. *Journal of Huntington's Disease, 6*, 63–72.

Van de Winckel, A., Feys, H., De Weerdt, W., & Dom, R. (2004). Cognitive and behavioural effects of music-based exercises in patients with dementia. *Clinical Rehabilitation, 18*(3), 253–260. https://doi.org/10.1191/0269215504cr750oa

van der Vleuten, M., Visser, A., & Meeuwesen, L. (2012). The contribution of intimate live music performances to the quality of life for persons with dementia. *Patient Education and Counseling, 89*(3), 484–488. https://doi.org/10.1016/j.pec.2012.05.012

Zheng, L., Vinters, H. V., Mack, W.J., Weiner, M. W., & Chui, H. C. (2016). Differential effects of ischemic vascular disease and Alzheimer's disease on brain atrophy and cognition. *Journal Cerebral Blood Flow Metabolism, 36*, 204–215.

# 13

# Therapeutic Music Interventions to Support People With Dementia Living at Home With Their Family Caregivers

*Jeanette Tamplin and Imogen N. Clark*

## Music Therapy in Dementia Care

Music therapists have observed and utilized the powerful affordances of music for people living with dementia for many decades. Music-therapy pioneers Ruth Bright and Alicia Anne Claire have been describing the benefits of music therapy in dementia care since the 1970s (Bright, 1973; Clair & Memmott, 2008). More recently, Hanne Mette Ridder has contributed significantly to this knowledge base, describing how music therapy and music interventions for people living with dementia in residential care facilitate human connection and effective communication (Ridder, 2011; Ridder & Gummesen, 2015; Ridder, Stige, Qvale, & Gold, 2013; Ridder & Wheeler, 2015). Ridder's research is underpinned by Kitwood's model of personhood, or person-centered care, which affirms the need for respectful trusting relationships for optimal well-being in dementia care (Kitwood & Bredin, 1992).

Interventions facilitated by qualified music therapists may include improvisation (spontaneous music creation), reminiscence-based song singing (singing songs that aim to stimulate autobiographical memories), supported songwriting (therapeutically informed creation of original songs), and receptive music listening (recipient of the music experience rather than active music making). These therapeutic music interventions offer unique and effective modes for social connection and engagement between care recipients and providers, as well as experiences of success and reduced isolation for people with dementia (Ridder & Wheeler, 2015). It is likely that these benefits seen in music therapy acted as mechanisms to decrease agitation in a recent randomized controlled trial (Ridder et al., 2013). Further evidence for the benefits of music therapy in dementia care is provided in a recent Cochrane review (van der Steen et al., 2017) and a systematic review with narrative synthesis (McDermott, Crellin, Ridder, & Orrell, 2013). The Cochrane review indicated that music therapy can decrease depressive symptoms in dementia and found low-quality evidence for improvements in emotional well-being, quality of life, behavioral problems, and cognition (van der Steen et al., 2017). The narrative synthesis review found consistent evidence for short-term improvements in mood and behavior (McDermott et al., 2013).

In keeping with the majority of music therapy research outlined above, music therapists have traditionally worked in aged care facilities, offering music-based interventions for people with mid to advanced stages of dementia (see earlier chapters in this book by Ridder, Mercadal-Brotons, Lipe, & Edmonston). However, worldwide, the majority of people with dementia live in their own homes rather than in residential care (Alzheimer's Disease International, 2015). This is the ideal place for people with dementia to reside for a number of reasons. The rapidly increasing dementia population means that it is not economically viable from a societal perspective to care for large numbers of people with dementia in aged care facilities, neither is it the ideal option in terms of care provision. The familiarity of the family home environment, where the person with dementia is surrounded by family members, well-known objects, and significant memories, is optimal for a person who is beginning to struggle with memory and orientation. For these reasons, there is a growing global trend toward supporting people with dementia (and their family caregivers) to live for as long as possible in the home of their choice (WHO, 2017). While such policies are common internationally, home care service providers still tend to focus primarily on physical needs and overlook psychosocial needs (Hansen, Hauge, & Bergland, 2017). There is clearly a need for community-based programs that address the psychosocial needs of people with dementia and support them to remain in their own homes with the support of family caregivers for as long as they wish to. Music therapy represents an ideal medium through which to target these outcomes.

## Music Therapy in Community-Based Dementia Care

In accordance with recent global trends and increasing numbers of people with dementia living at home, music therapists are now offering services to community-dwelling people with dementia and their families. Services range from early intervention to music-therapy community programs for mid to late stages of dementia, aiming to support people to live at home for as long as possible. Early intervention music-therapy programs can provide education about and access to music resources that the person living with dementia and their families can learn to use effectively to address specific dementia symptoms (or unmet needs), such as anxiety, apathy, and agitation, while also supporting meaningful social connection.

Home-based music-therapy programs are often informed by the theoretical frameworks of community music therapy and person-centered care. Community music therapy focuses on working musically with people in the context of the social and cultural factors of their health, illness, relationships, and music (Ansdell, 2002; O'Grady & McFerran, 2007). Sitting outside of the restrictions of traditional health care systems, community music therapy focuses on empowering people through participatory approaches to music making in inclusive, nonclinical settings (Stige, 2003). In dementia care, community music-therapy programs aim to use the normalizing and equalizing properties of music participation to maximize opportunities for musical expression and participation. The therapeutic effects of music participation that exist for everyone also apply to people living with dementia who retain the capacity to process and participate in music until late in the disease trajectory (Baird & Samson, 2009; Baird & Samson, 2015). Such benefits reported from music participation include positive effects on mood and well-being, increases in energy level and physical exercise; cognitive stimulation; collective bonding and peer support; a sense of contributing to something bigger and aesthetically beautiful; increases in confidence, motivation and self-esteem; and a sense of personal transcendence (Clift, Hancox, Staricoff, & Whitmore, 2008; Davidson & Garrido, 2015).

Despite the inevitable decline in function and increased need for support as the disease progresses, people with dementia are able to maintain their sense of self through autobiographical and musical memories (Baird & Thompson, 2018), social relationships (Desai, Desai, McFadden, & Grossberg, 2016; Johnson, 2016) and experiences of autonomy (Johnson, 2016). Opportunities for people living with dementia to collaborate as partners in therapy and maintain social connections are therefore essential. This is especially important in light of the high risk of social isolation for people living with dementia (Desai et al., 2016). The decline in cognitive function, in combination with the neuropsychiatric symptoms of dementia, can restrict access to social and community activities. Compounding this is the stigma surrounding a dementia diagnosis, which can lead to withdrawal from pre-morbid social groups (Harris & Caporella, 2014). People with dementia often report that as their disease progresses they become unable to continue to participate in their previous community music group or choir (Clark, Tamplin, & Baker, 2018). This can be due to cognitive decline impacting the ability to process music notation or instructions, or a sense of embarrassment or social stigma surrounding their dementia diagnosis. Therefore, there is a great demand for inclusive and dementia-friendly music programs and groups to address the need for social participation and maintenance of sense of self and musical identity. Such music-based dementia programs, when delivered by a music therapist, offer the additional benefit of the music therapist's clinical training regarding both the symptoms and trajectory of the disease, as well as knowledge of how best to appropriate the affordances of music for therapeutic benefit, maximize health and well-being, and minimize risk of harm for people living with dementia (Clark et al., 2018).

## The Role of Family Caregivers in Dementia Care

Family caregivers are the unsung heroes in dementia care. The majority of people with dementia living in the community receive informal care from family caregivers, who typically provide more than 40 hours of care per week (Brooks, Ross, & Beattie, 2015). These family caregivers have a difficult journey of their own. As well as needing to provide practical support (e.g., with personal care and activities of daily living) as symptoms of dementia increase, they also have to deal with the changes they see occurring in their loved one, make difficult decisions about treatment options and finances, grieve the loss of their pre-dementia relationship, and potentially make adjustments in social activities. Providing this level of physical and emotional support can have substantial negative effects on the caregiver's physical and psychological well-being (Afram et al., 2014), and can lead to social isolation and even financial burden if caregivers need to stop work to care for their loved one (Brodaty & Donkin, 2009).

In addition to its negative effects on caregiver health and well-being, caregiver burden can also affect the ability to care for their loved one living with dementia, thus leading to earlier admission to residential aged care. Family caregivers are clearly an essential part of the home-based dementia care equation and need to be supported in their role. A meta-analysis of interventions for caregivers of people with dementia (Brodaty, Green, & Koschera, 2003) found that psychosocial interventions such as counselling, support groups, stress management training, and education were effective in reducing caregiver distress. Such interventions were particularly effective if they involved both caregiver and person with dementia, with some studies reporting a delay in admission to residential care for participants engaged in such programs (Brodaty et al., 2003). Involving both caregiver and person with dementia in treatment planning also results in better outcomes for both (Boltz, Chippendale, Resnick, & Galvin, 2015).

The complexities of dealing with dementia symptoms together with the time-consuming nature and social stigma of caring for a loved one with dementia can also lead to increased risk for social isolation for family caregivers (Greenblat, 2012; Nay et al., 2015). To minimize the risk of social isolation for both person with dementia and family caregiver, we need programs for both to attend that focus on partnership and shared identity rather than individual needs only (Nay et al., 2015; Spector, Orrell, Charlesworth, & Marston, 2016; Wadham, Simpson, Rust, & Murray, 2016). In particular, strengths-based rather than deficit-orientated psychosocial services and interventions are particularly recommended to support relationships (Hellström, Nolan, & Lundh, 2005; Kitwood & Bredin, 1992; Merrick, Camic, & O'Shaughnessy, 2016; Spector et al., 2016; Wadham et al., 2016). Clearly, there is a need for psychosocial interventions designed to address the needs of the caregiver/care recipient dyad. Such dual-focused interventions may also be more cost effective and time efficient.

## Community-Based Therapeutic Singing Groups in Dementia Care: The Remini-Sing Model

Singing is one of the most potent tools for facilitating communication and social interactions in people with dementia (Ridder, 2011; Robertson-Gillam, 2011). The use of the voice when singing has a powerful connection to self, and even more so when the capacity to communicate verbally has deteriorated. In the later stages of dementia people lose the ability to speak, but many can still sing (Baird & Thompson, 2018). Singing is therefore one of the most effective ways to engage and build a connection with others. Research on group singing-based music-therapy interventions in residential dementia care has indicated improvements in confidence, motivation, communication, and social interaction (Dassa & Amir, 2014; Ridder & Aldridge, 2005; Ridder et al., 2013; Robertson-Gillam, 2008), as well as decreases in depression (Werner, Wosch, & Gold, 2015). In addition to reducing social isolation for people with dementia and their caregivers, intergenerational choirs have been used to decrease stigma and increase understanding of dementia and recognition of residual strengths and capabilities (Harris & Caporella, 2014).

Community-based singing groups for people with dementia or dementia-friendly choirs have been growing in popularity, particularly over the past decade. Such groups fall somewhere along the continuum of music experiences between community music and community music therapy (as defined in Chapter 2 by Clementes-Cortés) depending on facilitator, purpose, and philosophy, and differ from therapeutic singing groups facilitated by qualified music therapists. "Singing for the Brain" is an example of a community music program for people with dementia in the UK (Bannan & Montgomery-Smith, 2008; Osman, Tischler, & Schneider, 2016; Ward & Parkes, 2017). In Australia, community singing groups for people with dementia have also been examined in research studies (Davidson & Almeida, 2014; Davidson & Fedele, 2011). These models have generated qualitative evidence that community-based group singing is stimulating and accessible, normalizing (the focus is on singing rather than on dementia), socially engaging, and that it enhances relationship quality and improves emotional well-being (Camic, Myferi Williams, & Meeton, 2011; Unadkat, Camic, & Vella-Burrows, 2016). Only two studies have attempted to measure quantitative outcomes resulting from participation in community dementia singing groups and found no significant effects. Both studies used standardized measures to assess cognitive function and quality of life pre- and post-singing intervention (Camic et al., 2011; Davidson & Fedele, 2011). Camic and colleagues (2011) also measured mood, behavioral and psychological symptoms, and activities of daily living. The lack of effects from these standardized measures may indicate that these types of measures are either not sensitive enough or the singing program frequency (once weekly for 6 or 10 weeks) was not high enough to cause changes in the selected quantitative outcomes.

We have developed a community-based model of therapeutic group singing (Remini-Sing) for people living with dementia and their family caregivers (Tamplin, Clark, Lee, & Baker, 2018). Remini-Sing is process oriented, rather than product or performance oriented, which distinguishes it from dementia-friendly community choir models. As the name suggests, Remini-Sing focuses on singing familiar songs (suggested by participants) to stimulate autobiographical memory and social connection. Registered music therapist facilitators provide a therapeutic program utilizing variety of singing-based activities targeted to enhance memory, communication, well-being, and group cohesion. The model is based on a weekly 2-hour group consisting of opportunities for vocal warmup, singing familiar songs, learning new songs, simple harmony parts and rounds, social interaction, and peer support. Some of the practical elements of the Remini-Sing model include provision of name tags (to aid memory and social interaction), appropriate seating and positioning (some people need chairs with arms and it is important to arrange chairs so that participants can see one another), adequate space for mobility aids, accessible toilet facilities, and screen projection of lyrics rather than printed booklets (to avoid poor posture and confusion in finding song pages, and also to encourage participants to look up and at each other).

We conducted an inductive thematic analysis of interviews conducted with nine dyads who had participated in a Remini-Sing therapeutic singing group for 20 weeks (Clark et al., 2018). A number of interesting themes arose from this data analysis. Participants described the Remini-Sing group as making singing more accessible for people with dementia and their family caregivers. Some participants reported that they had been unable to continue to attend previous choirs due to cognitive demands, stigma, or lack of understanding of dementia by other choir members, or simply restrictions on time and freedom due to caregiving responsibilities. In the Remini-Sing group, they reported that there was a strong sense of acceptance and nonjudgment, that anyone could participate at whatever level they were able. The Remini-Sing group was also described as beneficial because the person with dementia and the caregiver could attend together and receive both individual benefits as well as a positive shared experience that didn't require a focus on care needs or additional respite support.

Singing together was a return to a past shared passion for some dyads and an opportunity to explore something new together for others. Singing in this therapeutic context was a "leveler," something that both people living with dementia and their caregivers could contribute to and do equally well (Clark et al., 2018). It was difficult to distinguish people with dementia from their caregivers in the singing group, and this was perceived as a positive and normalizing experience. This strengths-based approach meant that opportunities were provided for all participants to shine and flourish, and dementia became inconsequential to the group process. Participants also developed new supportive and empathic friendships through their participation in the Remini-Sing group. They felt connected not only in the music, but also in their shared experience of living with dementia and caring for a loved one with dementia. Participants described a feeling of forming "special bonds," which was particularly significant in the context of perceived increases in isolation due to lack of understanding by others and reduced opportunities for social participation elsewhere due to the effects of dementia. Ripple effects were also reported where the participants began to meet up and support each other outside of the weekly group singing sessions.

Participation in the Remini-Sing program also led to reported feelings of pride, increased confidence, a sense of purpose, and reconnection with self-identity (Clark et al., 2018). Being able to remember song lyrics and associated memories was reassuring for participants with dementia in the context of increasing memory difficulties in other areas of their life. Caregivers also observed benefits from the cognitive stimulation of the group, where their loved ones with dementia were able to learn new songs, remember the day of the week that the group was on, and initiate music participation at home following the group singing sessions. The weekly group singing sessions were experienced as engaging, enjoyable, and transportative, which has important implications for mood, emotional well-being, and quality of life (Clark et al., 2018). We are building on this feasibility research and currently conducting a randomized controlled trial to further evaluate the effects of the Remini-Song model. Based on these findings from the Remini-Sing project, we are also exploring the effects of songwriting for community-dwelling people with dementia and their families (Australian New Zealand Clinical Trials Registry: ACTRN12618000919213).

## Therapeutic Songwriting in Dementia Care

Therapeutic songwriting is a music-therapy method that has not traditionally been used in dementia care but is gaining recognition as a supportive intervention for both people living with dementia and their family caregivers (Baker, 2015; Baker & Stretton-Smith, 2017). Songwriting involves verbal and musical exploration of thoughts, feelings, attitudes, and memories (Baker,

2015) and may be a particularly powerful intervention during the early to mid phases of dementia when caregivers and care recipients are adjusting to the diagnosis and learning strategies for coping. Songwriting also offers a potent way of giving voice to people living with dementia to share their experiences. This is significant as the voices of people living with dementia are largely missing from current discourse in dementia research (Baker & Stretton-Smith, 2017). To date, group songwriting has been examined with either family caregivers or participants with dementia separately. This is not surprising since approaches to songwriting vary depending on the needs and participants.

People with dementia have cognitive difficulties such as memory and executive impairment, including difficulty initiating ideas, processing abstract thoughts, or perseveration (fixation on one topic). Nevertheless, Baker (2015) suggests that music therapists can assist in overcoming these challenges with careful planning. For example, therapists can use cues to refresh memory from earlier sessions as a point of reference from which to move forward. Musical cues are a particularly effective method for facilitating autobiographical recall (Baird & Samson, 2015). The use of empathic probing and sensitive questioning is also recommended by Baker (2015) to combat cognitive impairments and support participation. If the creation of original songs becomes too demanding, the therapist may incorporate structured or activity-based songwriting methods such as filling in missing lyrics or song parody (lyric substitution) to encourage participation (Baker, 2015). Using known songs as a primer to songwriting at the beginning of sessions may also be a helpful way to encourage participation, expression, and memory recall.

Recent studies have demonstrated cognitive, psycho-emotional, social, and spiritual benefits from songwriting for people living with dementia. Hong and Choi (2011) conducted a randomized controlled trial to compare effects of songwriting and usual care on the cognitive function of people with dementia in residential care (n = 30). Participants in the experimental music-therapy group received 16 weekly sessions involving preferred songs as a primer, structured songwriting methods, and performance of written songs. Between-group comparisons demonstrated significant improvements in language function, orientation and memory for the music-therapy group. In a case study, Ahessy (2017) described how individual songwriting was a validating experience for an elderly woman diagnosed with dementia who was living in residential care. Ahessy (2017) incorporated various types of songwriting from the development of completely original songs to more structured re-creations of known songs. The songs acknowledged the woman's personal history and also provided a means for emotional expression and validation of her repeated commentary. The woman claimed to recognize the songs from week to week, possibly because her own commonly used words and memories were embedded in the songs, and she was able to sing some of the new lyrics from memory. This case study also recognizes palliative aspects of dementia, noting that songwriting supports the expression of spiritual concepts associated with death and dying. Baker and Stretton-Smith (2017) examined songwriting sessions for four participants with early-stage dementia attending a community day care center. Semi-structured interviews both with staff at the center and the participants suggested that the sessions motivated participants, highlighted abilities, and led to feelings of accomplishment from the participants. Songwriting sessions were also perceived to facilitate social connections and stimulate creative and cognitive engagement. Day care center staff explained how their previous assumptions about what people with dementia can achieve were challenged through their observations of the therapeutic songwriting process.

## Songwriting for Caregivers of People Living With Dementia

Group songwriting sessions with caregivers of people with similar conditions, such as dementia, provide a safe and trusting forum where commonly experienced triumphs and challenges can be explored (Baker, Stretton-Smith, Clark, Tamplin, & Lee, 2018). Sessions over several weeks generally aim to address issues related to burden of care, adjustment to the caregiver role, altered family roles, and changes in the caregiver-recipient relationship. Other areas addressed include the caregiver's self-identity, grief and loss, anxiety related to an uncertain future, self-care, well-being, and skill development (Baker, 2017). Caregivers are encouraged and supported to contribute individual ideas for lyrics based on their experience of caregiving and discuss similarities and differences with other group members. No prior musical experience is necessary as the music therapist provides support and structure for the group to generate and collaboratively decide on musical elements for their song. With the emergence of initial lyric ideas, which are then refined and expressed with music, caregivers become involved in a therapeutic group process as they strive together towards the creation of a song. The process is cyclical with the initial generation of ideas, processing of ideas, and then reprocessing at a deeper level, thereby encouraging therapeutic expression from superficial to more complex issues and limiting opportunities for avoidance. The combination of both music and lyrics heightens meaning: for example, the musical attributes may be in conflict with lyrics, suggesting hidden tensions. Once completed, the song becomes a record of the therapeutic process, providing a deeply cathartic experience that can be repeated with each playing. In this way, the song becomes a reflection of the caregiver journey and allows songwriters to view their experiences from a new angle or perspective, thereby creating an alternative perception of life (Baker, 2015).

Baker (2017) describes a songwriting protocol for caregivers of people living with dementia based on problem-focused behavioral and emotion-focused cognitive theories. Three different approaches of songwriting are suggested within these frameworks: (a) "emotional-cognitive": insight-orientated songwriting, where the therapist directs clients to revise and re-frame their thinking and feelings in an attempt to overcome any barriers that are threatening health and well-being; (b) "problem-focused/behavioral": narrative (story-telling) songwriting which attempts to help people view issues from different perspectives, reshape negative thinking patterns, and form a more positive identity; and (c) "problem-focused/behavioral": psychoeducational songwriting with a focus on day-to-day stressors and reminders of pragmatic actions to mitigate stress. Baker (2017) suggests implementing these three approaches within a 12-week therapeutic songwriting protocol to create three songs using each of the different approaches. In weeks 1 to 4, caregivers are encouraged to express negative experiences of caring and reframe self-defeating thought patterns using an insight-oriented approach. In weeks 5 to 8 a narrative approach is used to explore the caregiver's altered current roles and how this differs from their previous lifestyle and identity. Finally, in weeks 9 to 12, a psychoeducational approach focuses on healthy coping strategies and skills is articulated in an effort to support ongoing adjustment over the longer term.

While Baker (2017) writes about a songwriting protocol involving creation of three songs in a systematic process involving several sessions, she also acknowledges that less intensive songwriting programs may also offer benefits to family caregivers. Klein and Silverman (2012) used psychoeducational approaches to compare single sessions of songwriting and discussion. Both the songwriting and discussion sessions incorporated the same guiding questions with a focus on self-care strategies. Thematic analysis found many similarities in participant responses to songwriting and traditional psychoeducational methods for family caregivers of people living with dementia in distracting them from stress, re-enforcing coping strategies, and improving insight regarding self and caring needs. However, only participants in the songwriting group remarked that sessions were fun, while there were no comments indicating such enjoyment from participants in the traditional discussion group.

Baker (2017) queried whether homogenous caregiver groups (e.g., all spousal caregivers) might create a more empathic songwriting environment than groups made up of people with different caregiver relationships (e.g., parent/son or daughter). To examine this notion, Baker and Yeates (2017) explored a four-session therapeutic songwriting program with four caregivers of people living with dementia. The caregivers were a male spouse, female spouse, son, and daughter. Interpretative phenomenological analysis of post-program interviews was used to explore caregivers' experiences of the therapeutic group songwriting process. Despite the heterogeneity of the group, participants described how working together with others in the group who had experienced similar circumstances and situations was meaningful both personally and for the group as an entity. Participants who were initially concerned about their perceived lack of musical ability found that the empathic environments along with support from the music therapist facilitated free expression and the creation of meaningful songs which exceeded their expectations. Despite the differing caregiver relationships, members of this group felt that the songwriting process helped them to learn more about themselves and each other, and to express the lived experience of the caring journey from various perspectives.

## Therapeutic Group Songwriting for People With Dementia Together With Their Family Caregivers

As discussed above, there is an increasing call for psycho-social interventions to focus on shared caregiver-recipient identity and capacity (Spector et al., 2016), particularly in the case of spousal or partner relationships (Hellström et al., 2005; Merrick et al., 2016; Molyneaux, Butchard, Simpson, & Murray, 2012; Wadham et al., 2016). To date, there appears to be no research examining group songwriting for dyads of people living with dementia together with their family caregiver. For couples who have been united for many years with shared memories and history, opportunities for them to participate in songwriting together may offer an opportunity to reminisce and explore shared identity, both present and past, thereby acknowledging and honoring strengths and resilience over the duration of their partnership. Songwriting might also offer couples and dyads living with dementia an opportunity to work through and share challenging issues together, such as changes in roles and responsibilities or intimacy. Since narrative therapy is recognized as particularly beneficial for couples receiving dementia care (Scherrer, Ingersoll-Dayton, & Spence, 2014), it would be interesting to explore the effects and experiences of songwriting for couples/dyads living with dementia, whether in groups or as family units.

## Caregiver-Implemented Music Interventions Guided by Music Therapy

With increasing numbers of people with dementia residing in their family homes, informal caregivers are now seeking interventions to assist with management of neuropsychiatric symptoms of dementia displayed by their loved one. There is a growing demand for caregiver-implemented music interventions (as distinct from formal music therapy) to support psychological well-being for both caregiver and care recipient (Särkämö, 2017). Music therapists are ideally placed to educate caregivers on enhanced communication strategies in dementia due to their expertise in engaging with clients in nonverbal and meaningful ways (Beer, 2017). Music therapists utilize musical tools, such as intonation, rhythm, and simplified speech and song to connect with people who are communicatively and socially isolated by their disease. Based on this knowledge, Beer (2017) developed and tested an enhanced communication training module for caregivers of people living with dementia focusing on attention to tone of voice, rhythm, and melody, and nuances of gesture as skills that can be taught to caregivers.

Research has shown that when a nurse caregiver sings during activities of daily living in dementia care (such as washing or feeding) agitation and violent behaviors decrease (Brown, Götell, & Ekman, 2001). Dyadic singing interventions as a part of everyday care implemented by family caregivers at home have also led to significant improvements in mood, orientation, and episodic memory in comparison to usual care for people living with dementia

(Särkämö et al., 2014). Interestingly, improvements were also found in short-term memory, working memory, and well-being for the family caregivers (Särkämö et al., 2014).

According to social exchange theory (Homans, 1961), role changes within caregiver-care recipient relationships lead to diminished reciprocity between a family caregiver and their loved one with dementia, creating imbalances in the relationship, and reducing intimacy. When music facilitates moment-to-moment interaction, emotional and social engagement, and autobiographical recall, there is an opportunity to address imbalances in reciprocity, particularly for spousal caregiver relationships (Baker, Grocke, & Pachana, 2012). Caregiver-implemented music interventions that are guided by music therapy offer great potential as engaging and cost-effective nonpharmacological approaches, with potential to improve the quality of informal care and mental health management in dementia care. A family caregiver is best placed to provide expert knowledge of their love one's musical preferences and background. They can utilize this knowledge, together with education provided by music therapists on how best to appropriate the affordances of this music, to manage behavioral and neuropsychiatric symptoms of dementia. A study by Baker et al. (2012) tested a home-based music intervention designed to stimulate meaningful interaction between ageing couples where one spouse has dementia. They found that the music-sharing experiences were beneficial to the relationship, with perceived increases in satisfaction with caregiving, caregiver well-being, and enhanced mood for both partners.

While music-therapy interventions in dementia care clearly offer great benefits, dementia prevalence is growing exponentially and music therapists are unable to meet the growing demand for services in this sector. Furthermore, it is likely that not all people living with dementia (particularly in the early stages of the disease) require the specialized skills of a music therapist. In consideration of these factors, together with the studies cited above, qualified music therapists need to develop training programs to assist informal caregivers to use music to support their caregiving role and to upskill community musicians to facilitate recreational group singing and music sessions.

## Conclusion

As the aging population rapidly increases, there is a subsequent growing incidence of dementia and an international trend for people with dementia to live longer in their own homes rather than in residential care (Alzheimer's Disease International, 2015). Therefore, music therapists need to develop community-based models to supplement traditional music-therapy models used in residential care facilities. Further, as there are not enough music therapists to fill the demand for music-therapy services in dementia care, music therapists need to develop consultancy models and training packages for caregivers and community musicians. Music interventions need to be more accessible during early stages of dementia, as a strengths-based resource to draw on through the full trajectory of the disease rather than focusing predominantly on symptom management in the later stages of dementia. Finally, music therapists clearly have a vital role to play in educating informal caregivers in how to best utilize music in their caring role to draw upon the strengths and capabilities of their loved one with dementia. Caregivers can also be empowered to use music as a resource to improve well-being, relationships, and quality of life, both for themselves and for the person they care for.

# References

Afram, B., Stephan, A., Verbeek, H., Bleijlevens, M. H. C., Suhonen, R., Sutcliffe, C., . . . Hamers, J. P. H. (2014). Reasons for institutionalization of people with dementia: Informal caregiver reports from 8 European countries. *Journal of the American Medical Directors Association, 15*(2), 108–116. doi:10.1016/j.jamda.2013.09.012

Ahessy, B. (2017). Song writing with clients who have dementia: A case study. *The Arts in Psychotherapy, 55*, 23–31. doi:10.1016/j.aip 2017.03.002

Alzheimer's Disease International. (2015). World Alzheimer report 2015—The global impact of dementia: An analysis of prevalence, incidence, cost and trends. Retrieved from https://www.alz.co.uk/research/WorldAlzheimerReport2015.pdf.

Ansdell, G. (2002). Community music therapy and the winds of change. *Voices: A World Forum for Music Therapy, 2*(2). doi:10.15845/voices.v2i2.83

Baird, A., & Samson, S. (2009). Memory for music in Alzheimer's Disease: Unforgettable? *Neuropsychology Review, 19*(1), 85–101.

Baird, A., & Samson, S. (2015). Music and dementia. In E. Altenmüler, S. Finger, & F. Boller (Eds.), *Music, neurology, and neuroscience: Evolution, the musical brain, medical conditions, and therapies* (Vol. 217, pp. 207–235.). Progress in Brain Research. Oxford: Elsevier.

Baird, A., & Thompson, W. F. (2018). The impact of music on the self in dementia. *Journal of Alzheimer's Disease, 61*(3), 827–841. doi 10.3233/JAD-170737

Baker, F. A. (2015). *Therapeutic songwriting: Developments in theory, methods, and practice.* New York: Palgrave Macmillan.

Baker, F. A. (2017). A theoretical framework and group therapeutic songwriting protocol designed to address burden of care, coping, identity, and wellbeing in caregivers of people living with dementia. *Australian Journal of Music Therapy, 28*, 16–33.

Baker, F. A., Grocke, D., & Pachana, N. A. (2012). Connecting through music: A study of a spousal caregiver-directed music intervention designed to prolong fulfilling relationships in couples where one person has dementia. *Australian Journal of Music Therapy, 23*, 4–21.

Baker, F. A., & Stretton-Smith, P. A. (2017). Group therapeutic songwriting and dementia: Exploring the perspectives of participants through interpretative phenomenological analysis. *Music Therapy Perspectives, 36*(1), 50–66. doi:https://doi.org/10.1093/mtp/mix016

Baker, F. A., Stretton-Smith, P., Clark, I. N., Tamplin, J., & Lee, Y. C. (2018). A group therapeutic songwriting intervention for family caregivers of people living with dementia: A feasibility study with thematic analysis. *Frontiers in Medicine, 5*, 151. doi:10.3389/fmed.2018.00151

Baker, F. A., & Yeates, S. (2017). Carers' experiences of group therapeutic songwriting: An interpretive phenomenological analysis. *British Journal of Music Therapy, 32*(1), 8–17. doi:/10.1177/1359457517728914

Bannan, N., & Montgomery-Smith, C. (2008). "Singing for the Brain": Reflections on the human capacity for music arising from a pilot study of group singing with Alzheimer's patients. *Journal of the Royal Society for the Promotion of Health, 128*(2), 72–78. doi:10.1177/1466424007087807

Beer, L. E. (2017). The role of the music therapist in training caregivers of people who have advanced dementia. *Nordic Journal of Music Therapy, 26*(2), 185–199. doi:10.1080/08098131.2016.1186109

Boltz, M., Chippendale, T., Resnick, B., & Galvin, J. E. (2015). Testing family-centered, function-focused care in hospitalized persons with dementia. *Neurodegenerative Disease Management, 5*(3), 203–215. doi:10.2217/nmt.15.10

Bright, R. (1973). *Music in geriatric care.* New York: St. Martin's Press.

Brodaty, H., & Donkin, M. (2009). Family caregivers of people with dementia. *Dialogues in Clinical Neuroscience, 11*(2), 217–228.

Brodaty, H., Green, A., & Koschera, A. (2003). Meta-analysis of psychosocial interventions for caregivers of people with dementia. *Journal of the American Geriatrics Society, 51*(5), 657–664. doi:0.1034/j.1600-0579.2003.00210.x

Brooks, D., Ross, C., & Beattie, E. (2015). *Caring for someone with dementia: The economic, social and health impacts of caring and evidence based supports for carers—Paper 42.* Brisbane: Alzheimer's Australia.

Brown, S., Götell, E., & Ekman, S. L. (2001). "Music-therapeutic caregiving": The necessity of active music-making in clinical care. *The Arts in Psychotherapy, 28*(2), 125–135.

Camic, P. M., Myferi Williams, C., & Meeton, F. (2011). Does a "Singing Together Group" improve the quality of life of people with a dementia and their carers? A pilot evaluation study. *Dementia, 12*(2), 157–176. doi:10.1177/1471301211422761

Clair, A. A., & Memmott, J. (2008). *Therapeutic uses of music with older adults.* Silver Spring, MD: American Music Therapy Association.

Clark, I. N., Tamplin, J., & Baker, F. A. (2018). Community-dwelling people living with dementia and their family caregivers experience enhanced relationships and feelings of well-being following therapeutic group singing: A qualitative thematic analysis. *Frontiers in Psychology, 9*, 1332. doi:10.3389/fpsyg.2018.01332

Clift, S., Hancox, G., Staricoff, R., & Whitmore, C. (2008). *Singing and health: Summary of a systematic mapping and review of non-clinical research.* Canterbury, UK: Sidney De Haan Research Centre for Arts and Health.

Dassa, A., & Amir, D. (2014). The role of singing familiar songs in encouraging conversation among people with middle to late stage Alzheimer's Disease. *Journal of Music Therapy, 51*(2), 131–153. doi:10.1093/jmt/thu007

Davidson, J. W., & Almeida, R. A. (2014). An exploratory study of the impact of group singing activities on lucidity, energy, focus, mood and relaxation for persons with dementia and their caregivers. *Psychology of Well-Being, 4*(1), 1–13. doi:10.1186/s13612-014-0024-5

Davidson, J. W., & Fedele, J. (2011). Investigating group singing activity with people with dementia and their caregivers: Problems and positive prospects. *Musicae Scientiae, 15*(3), 402–422. doi:10.1177/1029864911410954

Davidson, J. W., & Garrido, S. (2015). Singing and psychological needs. In G. Welsh, D. Howard, & J. Nix (Eds.), *The Oxford Handbook of Singing*. Oxford: Oxford University Press.

Desai, A. K., Desai, F. G., McFadden, S., & Grossberg, G. T. (2016). Experiences and perspectives of persons with dementia. In M. Boltz & J. E. Galvin (Eds.), *Dementia care: An evidence-based approach* (pp. 151–166). New York: Springer.

Greenblat, C. (2012). Dementia caregiving and caregivers. In D. Bramley (Ed.), *Dementia: A public health priority* (pp. 67–80). Geneva: World Health Organisation.

Hansen, A., Hauge, S., & Bergland, Å. (2017) Meeting psychosocial needs for persons with dementia in home care services—a qualitative study of different perceptions and practices among health care providers. *BMC Geriatrics, 17*(1). doi:10.1186/s12877-017-0612-3

Harris, P. B., & Caporella, C. A. (2014). An intergenerational choir formed to lessen Alzheimer's Disease stigma in college students and decrease the social isolation of people with Alzheimer's Disease and their family members: A pilot study. *American Journal of Alzheimer's Disease and Other Dementias, 29*(3), 270–281. doi:10.1177/1533317513517044

Hellström, I., Nolan, M., & Lundh, U. (2005) "We do things together": A case study of "couplehood" in dementia. *Dementia, 4*(1), 7–22. doi:10.1177/1471301205049188

Homans, G. (1961). *Social behavior: Its elementary forms*. New York: Harcourt Brace Jovanovich.

Hong, I. S., & Choi, M. J. (2011). Songwriting oriented activities improve the cognitive functions of the aged with dementia. *The Arts in Psychotherapy, 38*(4), 221–228. doi:10.1016/j.aip.2011.07.002

Johnson, H. F. (2016). Exploring the lived experience of people with dementia through interpretative phenomenological analysis. *The Qualitative Report, 21*(4), 695–711.

Kitwood, T., & Bredin, K. (1992). Towards a theory of dementia care: Personhood and well-being. *Ageing & Society, 12*, 269–287. doi:10.1017/S0`44686X0000502X

Klein, C. M., & Silverman, M. J. (2012). With love from me to me: Using songwriting to teach coping skills to caregivers of those with Alzheimer's and other dementias. *Journal of Creativity in Mental Health, 7*(2), 153–164. doi:10.1016/j.aip.2004.02.002

McDermott, O., Crellin, N., Ridder, H. M., & Orrell, M. (2013). Music therapy in dementia: A narrative synthesis systematic review. *International Journal of Geriatric Psychiatry, 28*(8), 781–794. doi:10.1002/gps.3895

Merrick, K., Camic, P. M., & O'Shaughnessy, M. (2016). Couples constructing their experiences of dementia: A relational perspective. *Dementia, 15*(1), 34–50. doi:10.1177/1471301213513029

Molyneaux, V. J., Butchard, S., Simpson, J., & Murray, C. (2012). The co-construction of couplehood in dementia. *Dementia, 11*(4), 483–502. doi:10.1177/1471301211421070

Nay, R., Bauer, M., Fetherstonhaugh, D., Moyle, W., Tarzia, L., & McAuliffe, L. (2015). Social participation and family carers of people living with dementia in Australia. *Health & Social Care in the Community, 23*(5), 550–558. doi:10.1111/hsc.12163

O'Grady, L., & McFerran, K. (2007). Community music therapy and its relationship to community music: Where does it end? *Nordic Journal of Music Therapy, 16*(1), 14–26. doi: 10.1080/08098130709478170

Osman, S. E., Tischler, V., & Schneider, J. (2016). "Singing for the Brain": A qualitative study exploring the health and well-being benefits of singing for people with dementia and their carers. *Dementia, 15*(6), 1326–1339. doi:10.1177/1471301214556291

Ridder, H. M. O. (2011). How can singing in music therapy influence social engagement for people with dementia? Insights from polyvagal theory. In F. A. Baker & S. Uhlig (Eds.), *Voicework in music therapy: Research and practice* (pp. 130–146). London & Philadelphia: Jessica Kingsley.

Ridder, H. M. O., & Aldridge, D. (2005). Individual music therapy with persons with frontotemporal dementia: Singing dialogue. *Nordic Journal of Music Therapy, 14*(2), 91–106. doi:10.1080/08098130509478132

Ridder, H. M. O., & Gummesen, E. (2015). The use of extemporizing in music therapy to facilitate communication in a person with dementia: An explorative case study. *Australian Journal of Music Therapy, 26*, 6–29.

Ridder, H. M. O., Stige, B., Qvale, L. G., & Gold, C. (2013). Individual music therapy for agitation in dementia: An exploratory randomized controlled trial. *Aging Mental Health, 17*(6), 667–678. doi:10.1080/13607863.2013.790926

Ridder, H. M. O., & Wheeler, B. L. (2015). Music therapy for older adults. In B. L. Wheeler (Ed.), *Music Therapy Handbook* (pp. 367–378). New York & London: Guilford Press.

Robertson-Gillam, K. (2008). *The effects of singing in a choir compared with participating in a reminiscence group on reducing depression in people with dementia* (Doctoral dissertation, Western Sydney University). Retrieved from https://researchdirect.westernsydney.edu.au/islandora/object/uws:5378/datastream/PDF/view

Robertson-Gillam, K. (2011). Music therapy in dementia care. In H. Lee & T. Adams (Eds.), *Creative Approaches in Dementia Care* (pp. 91–109). London: Palgrave Macmillan.

Särkämö, T. (2017). Music for the ageing brain: Cognitive, emotional, social, and neural benefits of musical leisure activities in stroke and dementia. *Dementia, 17*(6), 670–685. doi:10.1177/1471301217729237

Särkämö, T., Tervaniemi, M. A., Laitinen, S., Numminen, A., Kurki, M., Johnson, J. K., & Rantanen, P. (2014). Cognitive, emotional, and social benefits of regular musical activities in early dementia: Randomized controlled study. *The Gerontologist, 54*(4), 634–650. doi:10.1093/geront/gnt100

Scherrer, K. S., Ingersoll-Dayton, B., & Spence, B. (2014). Constructing couples' stories: Narrative practice insights from a dyadic dementia intervention. *Clinical Social Work Journal, 42*(1), 90–100. doi:10.1007/s10615-013-0440-7

Spector, A., Orrell, M., Charlesworth, G., & Marston, L. (2016). Factors influencing the person-carer relationship in people with anxiety and dementia. *Aging & Mental Health, 20*(10), 1055–1062. doi:10.1080/13607863.2015.1063104

Stige, B. (2003). *Elaborations toward a notion of community music therapy.* Oslo: Unipub.

Tamplin, J., Clark, I. N., Lee, Y. C., & Baker, F. A. (2018). Remini-Sing: A feasibility study of therapeutic group singing to support relationship quality and wellbeing for community-dwelling people living with dementia and their family caregivers. *Frontiers in Medicine 5*, 245. doi:10.3389/fmed.2018.00245

Unadkat, S., Camic, P. M., & Vella-Burrows, T. (2016). Understanding the experience of group singing for couples where one partner has a diagnosis of dementia. *Gerontologist, 57*(3), 469–478. doi:10.1093/geront/gnv698

van der Steen, J. T., van Soest-Poortvliet, M. C., van der Wouden, J. C., Bruinsma, M. S., Scholten, R. J., & Vink, A. C. (2017). Music-based therapeutic interventions for people with dementia. *Cochrane Database of Systematic Reviews, Issue 5*, Art No.: D003477. doi: 10.1002/14651858.CD003477.pub3

Wadham, O., Simpson, J., Rust, J., & Murray, C. (2016). Couples' shared experiences of dementia: A meta-synthesis of the impact upon relationships and couplehood. *Aging & Mental Health, 20*(5), 463–473. doi:10.1080/13607863.2015.1023769

Ward, A. R., & Parkes, J. (2017). An evaluation of a "Singing for the Brain" pilot with people with a learning disability and memory problems or a dementia. *Dementia, 16*(3), 360–374. doi:10.1177/1471301215592539

Werner, J., Wosch, T., & Gold, C. (2015). Effectiveness of group music therapy versus recreational group singing for depressive symptoms of elderly nursing home residents: Pragmatic trial. *Aging & Mental Health, 21*(2), 147–155. doi:10.1080/13607863.2015.1093599

WHO. (2017). *Global plan of action on the public health response to dementia 2017–2025.* Geneva.

# 14

# Future Directions

Amee Baird, Sandra Garrido, and Jeanette Tamplin

> ". . . to those who are lost in dementia. . . Music is no luxury to them,
> but a necessity, and can have a power beyond anything else to restore
> them to themselves, and to others, at least for a while."
> —Oliver Sacks (2007)

There is no doubt that music is a special stimulus for people with dementia. As we have learned from each chapter in this volume, music can have a wide range of positive effects—it can help to access long forgotten memories, stimulate movement and speech, soothe depressed and anxious minds, and enhance relationships. In this way, it can be considered a viable frontline treatment for some of the most problematic symptoms and behaviors exhibited by people living with dementia. As this volume shows, the benefits of music for people with dementia are now well established, although there is a need for continued robust research in this field.

We must now move beyond merely describing these benefits, to further our understanding of how and for whom music is most beneficial. Before discussing the need for further research in these areas, there is one important issue that needs to be acknowledged, which is that music is not a cure for dementia. As described in Chapters 6 (Särkämo) and 8 (Baird and Thompson), music can alleviate certain symptoms, and serve as a crucial means of "access"—to emotions, memories, a sense of self, and other people—but it cannot reverse the inevitable decline associated with a neurodegenerative condition. Nevertheless, as an enjoyable stimulus for the majority of people, music interventions are cost-effective and highly accessible, and the risks associated with its use are minimal in comparison to pharmacological treatments. Music interventions are certainly a promising alternative to pharmacological treatments.

## For Whom is Music Beneficial?

As discussed in many chapters of this volume, most extant studies have focused exclusively on people with Alzheimer's Disease (AD). This is likely
due to the frequently observed positive responses to music for people
living with AD (as reported by Vanstone and Cuddy in Chapter 4), which
results from the relative preservation of memory for music and the subsequent potential to use music to increase engagement. Therefore, a clear
need exists for future research to develop our understanding of the effects
(if indeed there are any) of music on symptoms of non-AD dementias, such
as Parkinson's-related dementia, Lewy Body Dementia, and Huntington's
Disease. As discussed by Omar in Chapter 5, there is increasing understanding of the music cognition profiles of people with frontotemporal
dementia, particularly the behavior variant, but less so for the other
variants. We know that there are differences in music cognition and experience of people with behavioral variant fronto-temporal dementia
(Bv-FTD) compared with those with AD, including the common experience of "musicophilia" (the abnormally enhanced craving for music), and
reduced frequency of music-evoked autobiographical memories, as briefly
described by Baird and Thompson in Chapter 8. These differences are not
surprising given that the neuropathology of AD and Bv-FTD affects different brain regions (primarily frontal and temporal respectively). Other
dementia types such as Parkinson's-related and Lewy Body dementia primarily affect subcortical regions, which may result in specific music cognition deficits (e.g., deficits in rhythmic processing; see Grahn & Brett,
2009) and altered emotional and mnemonic responses to music compared
with those with AD, who have relative integrity of subcortical brain regions. There is clearly a gap in our current knowledge regarding the effects
of music in people with non-AD dementia, which needs to be addressed
by further research. In particular, there is often scarce attention paid to
different dementia types in the music-therapy literature and it is this field
that has the most promising potential for developing targeted treatments
for people with various dementia types. While a focus on differentiating
dementia types may present a methodological challenge for researchers
(given that some people with dementia may not have received a clear diagnosis specifying their sub-type), a greater understanding of differing
responses to music in the dementia sub-types will significantly enhance the
development of targeted music-based interventions.

A further consideration that is relevant to understanding for whom music is effective is the fact that music is certainly not a "one size fits all" treatment for people with dementia, despite commonly held misconceptions promoted in the popular media. It cannot be assumed that music will have positive effects for all people with dementia. In fact some may not experience any benefits from music at all, such as those who have neurological conditions that impair the processing of auditory stimuli or those who find music listening uncomfortable or agitating. Every individual has their own preferred music, their own history with music, an individual mental health record, and differing current symptoms. This means that often interventions using music that is not personally relevant to an individual return mixed results (Garrido et al., 2017). Furthermore, the way an individual has interacted and responded to music in the past is no guarantee that they will respond to music the same way in later stages of life, particularly once cognitive impairment sets in. Physiological changes to the ear associated with ageing also affect the type of music individuals are attracted to (Gates & Hills, 2005; Gordon-Salant & Frisina, 2010). Furthermore, the cognitive mediation of raw physiological responses to music, which usually occurs in healthy individuals (Konecni, Brown, & Wanic, 2008), may not occur to the same degree in people with cognitive impairment, causing responses to music to be much more instinctive. As demonstrated by Garrido in Chapter 7, individuals with a history of depression or other mood disorders may also be susceptible to negative responses to music, and music-based interventions therefore need to be tailored not only to the taste of the individual but also to their clinical symptoms. Thus, in addition to developing instruments for assessing the degree to which an individual is likely to respond to music, there is a need for further research to understand the types of music that provide optimal mood management for individuals of varying preferences, symptoms, and histories.

## How is Music Beneficial?

Several chapters in this book address the important question of *how* music has positive effects for people with dementia. In Chapter 3, Brancastisano and Thompson outline the specific capacities of music that make it uniquely effective as a treatment tool, and in Chapter 2 Ghilain and colleagues highlight the rhythmic entrainment effects of music. The potential for music to access various types of preserved memory functions in people with AD that cannot be expressed with other types of stimuli is highlighted by Baird and Thompson in Chapter 8. These chapters contribute to our growing understanding of the mechanisms that may underlie the beneficial effects of music. Nevertheless, there is still much to learn about the neural and functional mechanisms that underpin music's powerful effects. Furthermore, we need more methodologically rigorous research with appropriate control conditions to further understand if these effects are actually unique to music. Given the increasing consideration and use of music as a nonpharmacological alternative to address specific symptoms of dementia, it is imperative that there is a corresponding increase in our understanding of how music has its effects, in the same way that the effects of pharmacological treatments receive close research attention.

## Principles of Best Practice and Guidelines for Music Use in Dementia

Future research should inform and guide the development of principles of best practice for the use of music with people with dementia. There is growing awareness by the public of the positive effects that music can have for people with dementia, but many health care professionals and family caregivers are unsure how to implement this knowledge. Furthermore, we need to develop specific guidelines for how best to utilize the affordances of music at different stages of the dementia trajectory. Perhaps in the earlier stages of the disease, the focus of music interventions is more on social engagement and empowerment, and in the later stages the most powerful use of music is in affirming personhood and identifying unmet needs. Closely related to the development of guidelines is the need to understand the best ways to measure the impact of music participation for people with dementia, as described by Dowson and McDermott in Chapter 9. Once we can confidently measure and understand the effects of music participation and music-based interventions, we will be better able to translate this research into practical guidelines that can be utilized by people with dementia and caregivers living in the community or in aged care facilities.

As demonstrated by the systematic reviews presented in Chapters 10 (Lipe & Edmonston) and 12 (Mercadel-Brotons), it is clear that there is a growing evidence base for music therapy and music-based interventions in dementia care. There is also an increasing awareness of the differential benefits of music-based interventions (such as community singing) and receptive music activities (such as personalized playlists) to goal-based clinical music-therapy interventions (as described by Clements-Cortes in Chapter 1). It is important to understand these differential benefits as it will not be feasible for limited numbers of trained music therapists to meet the growing demand for music-based services in dementia care. Furthermore, as discussed in Chapter 13 by Tamplin and Clark, there are situations in which the specialized skills of a music therapist are not required, particularly in the earlier stages of the disease. Nevertheless, music therapists with their expertise in person-centered, nonverbal, and alternative communication methods are uniquely placed to provide helpful training and guidance to care providers. In Chapter 11, Ridder and Ørnholt Bøtker clearly demonstrate this through a detailed case presentation where they illustrate the skills of a trained music therapist in identifying and responding to musical cues that affirm personhood in advanced dementia. They then describe how attuned musical interactions can be transferred to daily care through skill sharing and inter-disciplinary collaboration.

There is now a need for the development of standardized procedures for healthcare professionals and aged care service providers to implement music activities. What works for a group-based activity in an aged care facility may not be effective or relevant for a couple living at home. Music has many affordances and can be a powerful medium for evoking memories and emotional responses (see Chapters 2, 3 and 6). For this reason, music also has the potential to stimulate negative responses or even cause harm (as discussed by Garrido in Chapter 7). If used without considering the physiological effects of musical elements and assessment of the individual circumstances and subjective responses to different types of music, it is possible that music may increase agitation, confusion, or emotional distress. Technological advances have made music more highly accessible that ever before. Together with well-meaning promotion of the benefits of music in dementia by popular media, these technological advances make it easier to offer music to people with dementia. The development of guidelines for music use in dementia and utilizing technology that enables easy access to music and dissemination of information is needed.

Finally, in addition to the potential benefits of music for people living with dementia, there is growing evidence to indicate a positive ripple effect of music on care providers. This experience of music may directly affect caregivers personally (such as mood enhancement) and/or have a flow on effect through decreasing stress (for example, if the person they care for is less agitated). When used appropriately and intentionally, music experiences can increase intimacy and social engagement and address reciprocity imbalances in relationships (Baker, Grocke, & Pachana, 2012). There is also vast potential to build support networks though community-based music-therapy programs, as identified by Tamplin and Clark in Chapter 13. Music is a positive reason to draw groups together and also a great "leveler" in terms of potential for all participants (with and without dementia) to contribute meaningfully.

This book has presented a multidisciplinary perspective on the effects of music engagement for people living with dementia. Experts from music psychology, music therapy, and clinical neuropsychology in the area of music and dementia have provided an overview of the status of research in this field to date. We hope that this book will both inspire and help to direct further research in this promising domain. We also hope that the information provided will guide the work of therapists, clinicians, care providers, and family members in how best to use music with people who are living with a dementia diagnosis. With increasing rates of dementia worldwide, and no cure in sight, music may be a universal source of hope for people with dementia and also for those caring for them.

Amee, Sandra, and Jeanette, December 2018

# References

Baker, F. A., Grocke, D., & Pachana, N. A. (2012). Connecting through music: A study of a spousal caregiver-directed music intervention designed to prolong fulfilling relationships in couples where one person has dementia. *Australian Journal of Music Therapy, 23*, 4–21.

Garrido, S., Dunne, L., Chang, E., Perz, J., Stevens, C., & Haertsch, M. (2017). The use of music playlists for people with dementia: A critical synthesis. *Journal of Alzheimer's Disease, 60*, 1129–1142. doi:10.3233/JAD-170612

Gates, G. A., & Hills, J. H. (2005). Presbycusis. *The Lancet, 366*, 1111–1120.

Gordon-Salant, S., & Frisina, R. D. (2010). Introduction and overview. In S. Gordon-Salant, R. D. Frisina, A. N. Popper, & R. R. Fay (Eds.), *The aging auditory system* (Vol. 34, pp. 1–8). New York: Springer.

Grahn, J. A., & Brett, M. (2009). Impairment of beat-based rhythm discrimination in Parkinsons' Disease. *Cortex, 45*(1), 54–61.

Konecni, V. J., Brown, A., & Wanic, R. A. (2008). Comparative effects of music and recalled life-events on emotional state. *Psychology of Music, 36*, 289–308.

# Index

Tables, figures, and boxes are indicated by an italic *t*, *f*, and *b*, respectively, following the page number.

*For the benefit of digital users, indexed terms that span two pages (e.g., 52–53) may, on occasion, appear on only one of those pages.*